Looking *for* Longleaf

Lawrence S. Earley

Looking *for* Longleaf

The
Fall
and
Rise
of an
American
Forest

The University of North Carolina Press
Chapel Hill and London

The author and the University of North Carolina
Press thank the Southeast Division of The Nature
Conservancy, the Longleaf Alliance, and the Tall Timbers
Research Station for their financial support of this book.
Each of these organizations is dedicated to the conservation
and restoration of the remaining examples of the longleaf
pine ecosystem across the southeastern United States and
believes that a broad public understanding of the history
and ecological diversity of that ecosystem will help
make its conservation a reality.

Designed by Heidi Perov
Set in Dante by Tseng Information Systems, Inc.
Manufactured in the United States of America

The paper in this book meets the guidelines for
permanence and durability of the Committee on
Production Guidelines for Book Longevity
of the Council on Library Resources.

Library of Congress Cataloging-in-Publication Data
Earley, Lawrence S.
Looking for Longleaf : the fall and rise of an
American forest / by Lawrence S. Earley.
p. cm.
Includes bibliographical references (p.).
ISBN 0-8078-2886-6 (cloth : alk. paper)
1. Longleaf pine — Southern States. 2. Forests
and forestry — Southern States. I. Title.
SD397.P59E27 2004
634.9'751'0975 — dc22
2004000724

08 07 06 05 04   5 4 3 2 1

Portions of this book appeared previously,
in somewhat different form, in the June 1997 and
August 2004 issues of *Wildlife in North Carolina*
magazine and in The Nature Conservancy, *Managing
the Forest and Trees: A Private Landowner's Guide to
Conservation Management of Longleaf Pine*, and
are reprinted here with permission.

To

Joseph Earley,

my father,

and to

Teresa Elizabeth

Earley,

my mother,

for their gifts

to me

Contents

Acknowledgments

In a very real sense this book is as much about the scientists and landowners who have devoted great portions of their lives to the study and growing of longleaf pine as it is about the longleaf pine ecosystem itself. Their dedicated work to understand and conserve the ecosystem is a story of research and applied science at their best. A great many of these people have guided me on my journey into history, forestry, and ecology, and as I am neither a historian, forester, or ecologist I want to acknowledge them all.

I would like to thank all whose names are mentioned in the pages of this book for taking the time to accompany me into the field and for patiently answering my numerous questions. Several others not mentioned helped me at various points in the writing. I am grateful to the North Carolina Wildlife Resources Commission for the opportunities I was given in working for *Wildlife in North Carolina* magazine over my twenty-year career there. A. Sidney Baynes, chief of the commission's Division of Conservation Education when I began this book in 1994, encouraged me in too many ways to mention. Jim Dean, my editor and friend at *Wildlife in North Carolina*, and Vic Venters, former assistant editor, listened to my half-coherent ramblings while I was writing early chapters of the book. I am indebted to all of them.

I owe a special debt to botanist Julie Moore, a friend to many conservation causes in the Southeast, especially that of the longleaf pine, and my mentor through all stages of manuscript preparation. This book began as an exploration into the forgotten history of turpentining, but I soon realized that this history would be told most effectively in the context of a book on the longleaf pine ecosystem. It was Julie's passion for the natural communities of the Southeast and their conservation that led me to broaden my work and gave me the grit to continue it when it would have been easier to drop such a daunting task. That I have completed the book at all may well be because I didn't want to tell Julie that I didn't.

I would like to acknowledge the assistance of the following individuals and their organizations in making the publication of this book possible: Nelwyn McInnis and Rob Sutter of The Nature Conservancy; Rhett Johnson and Dean Gjerstad of the Longleaf Alliance; and Lane Green, Ron Masters, and Lennie Brennan of Tall Timbers Research Station. Tall Timbers also provided me with lodging for several research stays and the use of their excellent library services. The conversation and guidance offered by the community of resident and visiting scientists at Tall Timbers were indispensable to my project. Auburn University provided welcome hospitality at the Solon Dixon Forestry Education Center in Andalusia, Alabama.

To the readers who have read all or parts of the manuscript during its long gestation, I can only offer my deepest appreciation. Jeffrey Richards read early versions of several chapters, as did Joan Berish, Steve Hall, Bob Peet, and Alan Weakley. Bill Platt read several drafts of chapters as well as the completed manuscript and offered invaluable advice. Catherine Bishir edited the manuscript fiercely and helped me pare down an unwieldy tome. Robert Farrar read a complete draft of the manuscript and took me to task for the errors and absences therein. The errors that remain, of course, are mine alone.

My debt to my wife Renee Gledhill-Earley for her loving patience is beyond repaying.

And to my young neighbors, Helen Johnston and Emme Johnston, I can now say: Yes, I have finished the book, and, no, you won't have to wait until you have children of your own to read it.

Looking *for* Longleaf

Land of the Longleaf Pine

Here's to the land of the longleaf pine
The summer land where the sun doth shine,
Where the weak grow strong and the strong grow great,
Here's to "down home," the Old North State!
—From the North Carolina state toast

The longleaf pine forests and savannas of the southeastern United States once comprised one of the most extensive woodland ecosystems in North America. Longleaf pine was the dominant tree over about 60 million acres of the Southeast when the Spanish arrived in the early 1500s, and it mixed with other pines and hardwoods on an additional 30 million acres. These great conifer forests sprawled over nearly 150,000 square miles, covering a wide swath of every coastal state from the James River in southeastern Virginia as far south as the shores of Lake Okeechobee in the Florida peninsula and west to southeastern Texas, interrupted only by the vast floodplain of the Mississippi River. Longleaf pine covered such a wealth of country that you could travel for months through it and feel as if you had never left home. Not long after the Revolutionary War, one settler moved from Emanuel County, in eastern Georgia, across the Creek Nation in present-day Alabama, to the Leaf River in eastern Mississippi where he built a house and raised cattle. The trip covered about four hundred miles and lasted four months, and he settled on the Leaf because it reminded him of Emanuel County. "The turpentine smell, the moan of the wind through the pine-trees, and nobody within fifty miles of him, was too captivating a concatenation to be resisted, and he rested here," his grandson said.[1]

Longleaf pine (*Pinus palustris*) still grows in a variety of settings. I've been surrounded by longleaf pines while listening to the Gulf of Mexico's gentle surf a few yards away, and I've walked through longleaf pine forests two thousand feet high in northern Alabama. I've seen stunted longleaf thriving, where few other plant species could, in ancient sand dunes, and I've understood why early travelers called them "barrens." I've wandered through sun-drenched, grass-covered longleaf savannas dotted with colorful meadow beauties, orchids, asters, and scores of other herbaceous species, and understood why travelers sometimes compared them to "meadows." Longleaf pine is described as mostly a tree of poor, sandy soils, but in the rich loams of Mississippi and Louisiana it grew phenomenally, and I've understood why these dense forests were called the "Piney Woods." Though mostly confined to the Coastal Plain, longleaf can also be found in sections of the Piedmont and, strikingly, in the southern foothills of the Appalachian Mountains.

The land of the longleaf pine is a land of great beauty but also of great violence, born from an encroaching sea and shaped by rivers, storm, and fire. It carries the evidence of immensities of time spent beneath ocean waters; the shells of sea creatures lie thick in the soil. Past seas shaped this land's terrain, rivers its very pitch and roll. It's a green land, yet also a land of flame and ash. It was in this region of sand and sea, wind and fire that longleaf flourished as no other pine could do, perfectly adapted to the conditions, swaggering over the Coastal Plain from southeastern Virginia to eastern Texas.

Yet little is left of the longleaf pine ecosystem today: only 1.4 percent of the Atlantic and Gulf Coastal Plains still support longleaf, compared to 60 percent in presettlement times. By 1996 only 2.95 million acres of longleaf remained out of the original 92 million acres, which may sound substantial until you realize that this acreage consists of fragments lost in the immensity of 150,000 square miles. Almost all of the old-growth forest is gone — perhaps 12,000 acres remain in scattered stands. By any measure, longleaf's decline of nearly 98 percent is among the most severe of any ecosystem on earth. It dwarfs the Amazon rain forest's losses of somewhere between 13 and 25 percent. It is comparable to or exceeds the decline in the North American tallgrass prairie, the coastal forests of southeastern Brazil, and the dry forests of the Pacific Coast of Central America. It surpasses the losses of the old-growth Douglas fir forests of the Pacific Northwest (87 percent in 1990).[2]

Conservation biologist Reed Noss has listed longleaf pine among the most seriously endangered ecosystems in the United States, and ecologist Cecil Frost has called the devastation of longleaf pine forests "a milestone event in the natural history of the eastern United States, at least equal in scale and impact to the elimination of chestnut from Appalachian forests by blight."[3]

Longleaf's decline has been attributed to a great many things but is most easily explained in three words—need, greed, and mismanagement. People cut the forest, burned it to farm and make spaces to live, exploited its resources, and changed the natural processes that had evolved with it and maintained it. The cast of characters includes farmers, industrial turpentiners, lumber companies, paper companies, foresters, and others. All of them in some way made their livings from the forest and tried to shape it for their own ends. Some of them nearly destroyed it to satisfy their wants and expressed no remorse when their work was done. One lumberman in 1893 forthrightly stated that the cut-and-run loggers then beginning to pillage the great bald cypress and longleaf forests throughout the South were entirely indifferent to their fate, and rightly so.[4]

In 1995 a trio of eminent longleaf specialists lamented that "what once was one of the most extensive forest ecosystems in North America [longleaf pine] has nearly vanished without notice." *Without notice?* That was a bit of an exaggeration. Longleaf's passage from the greatest conifer forest on the continent to a fragment has been long and torturous—death by a thousand cuts—but it's been especially well noticed and well mourned, a subject of scientific and even popular remark for more than 150 years. As early as the 1850s, industrial turpentiners had hacked up so many longleaf pine trees, leaving their bark coated with resin that dried white, that travelers often compared them to sheeted specters and predicted the forest's ultimate demise. "The time is not far distant when these stately monarchs of the forest . . . will have been borne down by the unwearied worker at their feet and not one vestige of their former glory will remain," wrote one traveler in North Carolina in 1854. "Aye well may you weep melancholy tree for your days are numbered."[5]

Seventy years later, North Carolina poet Anne McQueen was still echoing this sentimental refrain:

Listen! The great trees call to each other:
"Is it come your time to die, my brother?"
And through the forests, wailing and moaning,
The hearts of the pines, in their branches groaning:
   "We die, we die!

Flaying the bark, and our bodies baring,
Like dim, white ghosts in the moonlight staring,
Naked we stand, with the life-sap welling—
Tears of resin to gather for selling—
   We die, we die!"

All through the land are the forests dying,
One piece of silver a tree-life buying;
Listen! The great trees moan to each other:
"The ax has scarred us too, my brother—
    We die, we die!"[6]

The chorus of critical voices intensified in the twentieth century, when the wastelands of stumps left by cut-and-run loggers angered early conservationists in the same manner as the loss of endangered species and wetlands or the harvesting of the old-growth forests of the Pacific Northwest provoke conservationists of our own day. For North Carolina's eminent ecologist, B. W. Wells, longleaf's fate was an indictment of the entire capitalist system. "In its pristine condition with millions of trees measuring a yard or more in basal diameter, the [longleaf pine forest] unquestionably presented one of the most wonderful forests in the world," he wrote in 1931. "And today hardly an acre is left in North Carolina to give its citizens a conception of what nature had wrought in an earlier day. The complete destruction of this forest constitutes one of the major social crimes of American history."[7]

The conservation of longleaf pine as an issue and even a cause has never really disappeared. During the twentieth century scientists and conservationists wrote thousands of papers and held a score of symposia on the beleaguered forest. Several generations of research foresters have spent careers trying to understand the demands of the tree, and thousands of landowners have struggled to grow it despite its many difficulties.

In the 1980s, after decades of frustrating failures, the U.S. Forest Service pioneered new silvicultural techniques that made growing longleaf more reliable for the private grower. New muscle in environmental laws, especially in the application of the Endangered Species Act, pushed longleaf pine management to the forefront on many public lands in the South. At century's end, the restoration of this once great ecosystem was a major initiative throughout the South.

Saving longleaf has arguably been the longest-lived forest issue in American conservation history. Why this issue has proved so intractable is the subject of this book.

PART ONE

Ecology

What Bartram Saw

A magnificent grove of stately pines, succeeding to the expansive
wild plains we had a long time traversed, had a pleasant effect, rousing the
faculties of the mind, awakening the imagination by its sublimity, and
arresting every active, inquisitive idea, by the variety of the scenery.
—William Bartram, *Travels* (1791)

A longleaf pine forest on a bright day is a light and sound show. There's the verdant ground cover, mostly grasses that sway to each hint of breeze. The forest is open with widely scattered trees, and the early morning sun casts angled shadows from the pine trunks; by midday each tree will be standing in its own small pool of shadow. Here and there, dense groups of young pine saplings gather and the tufts of infant pines are nearly indistinguishable from the wiregrass. Above, the sky burns azure. The sound emanates from the treetops, a low and constant tone like the surf crash of a distant sea. Even on a perfectly still day you may hear this roar in the distance, as if somewhere an individual tree was gathering and amplifying some ambient sound. The great eighteenth-century explorer William Bartram described it as "the solemn symphony of the steady Western breezes, playing incessantly, rising and falling through the thick and wavy foliage."[1]

On a sunny morning in April, I've come to the 200-acre Wade Tract Preserve near Thomasville, Georgia, to walk through an old-growth longleaf pine forest. Old-growth longleaf pine is scattered in small pieces throughout the Southeast, unlike the Pacific Northwest where relatively large tracts of old-growth Douglas fir still exist. The Wade Tract is one of these remnant longleaf forests. It's owned by the Arcadia Plantation and managed, through a conservation easement, by Tall Timbers Research Station just down the road. This rolling country is known as the Red Hills region, where erosion over the eons has carved an originally flat plain into pleasant hills and valleys.

Some of the older longleaf pines have a distinct lean to them, and their tops have flattened with the loss of branches. Longleaf can grow to a ripe old age, about 400 to 500 years. The heights of the trees vary from 50 or 60 feet high in the deepest sands of the Carolina Sandhills to 110 feet or taller in richer soils. Their girth is modest—anything larger than 3 feet in diameter at breast height is really large; many old-growth trees had diameters of less than 2 feet measured at breast height. Longleaf is a beautiful tree, with lower branches that are undulant and graceful and that carry large cones. Its long needles distinguish it among all other pines and give it its name.

On this spring day, the red-headed woodpeckers are in frenzied motion, darting after each other among the pines and drumming incessantly on dead trees. They are mating and establishing territories, displaying the broad, black and white patterns of their wings, their large black bodies and crimson heads.

Grass is the predominant type of plant in this forest. There are possibly dozens of species growing here, although the most common is wiregrass (*Aristida stricta*). It grows green in spring and summer and turns a vivid gold in fall and winter, in all seasons rippling and bending in the wind. A wildfire ran through the forest three weeks ago, blackening some of the tree trunks and turning their needles a copper color. Yet the wiregrass has already greened out and pushed up into the sun, and the landscape looks scrubbed and fresh.

Across the rolling, parklike landscape of randomly spaced trees the open vista quickly thickens with distant trees. If I ambled off this path and through the wiregrass, past a drain that has thickened with a few shrubby oaks and up the sun-dappled hillside beyond, I'd see another vista just like the last. And then another.

I'm thinking: *Perhaps this is what Bartram saw.*

━━━━━━━

Not John Bartram, the famous Pennsylvania botanist to the King of England, friend of Benjamin Franklin, explorer and naturalist, but his son, William. Both Bartrams explored the southeastern United States in the late eighteenth century and wrote about their encounters with the longleaf pine forests. You can find John Bartram's account of their trip in a good research library, although it might prove skimpy reading. "Fine warm morning. Birds singing, fish jumping, and turkies gobbling," he said about one particularly fine day. John's friends and supporters shook their heads over his sketchy travel accounts. One noted that "he did not care to write down his numerous and useful observations. . . . He is rather backward in writing down what he knows."[2]

Not so William. The younger Bartram accompanied his father on his first journey to the Southeast concluding in 1766 and then, alone this time, covered almost the same itinerary beginning about seven years later. He had been commissioned by London physician and fellow Quaker, Dr. John Fothergill, to collect botanical specimens and make botanical drawings of his travels. From Pennsylvania, he sailed to Charleston and explored the region around Savannah, pushing up the Savannah River to Augusta before continuing south to Florida. He negotiated the St. Johns River by canoe, accompanied an expedition of Indian traders west across Florida, pressed into northern Georgia and the Carolina highlands in Cherokee country, and then made his final trip west to the Mississippi River. Intended to take two years, William's travels actually lasted five (1773–77). Throughout that time he was rarely out of sight of longleaf pine.

William Bartram's account of his trip, originally entitled *Travels through North and South Carolina, Georgia, East and West Florida, the Cherokee Country, the Extensive Territories of the Muscogulges, or Creek Confederacy, and the Country of the Chactaws*, published in 1791, provides one of the earliest and most detailed descriptions of the virgin longleaf pine forests, although his literary style takes a while to get used to it. He was a practitioner of the eighteenth-century literary school in which a noun without an adjective is like a man without his pants. Often he seems to overwhelm the scene he's describing with the artificial flowers of his prose: "At cool eve's approach, the sweet enchanting melody of the feathered songsters gradually cease, and they betake themselves to their leafy coverts for security and repose."[3] His father might have said, had he been tempted to say anything at all about such matters, "The birds stopped singing."

Bartram was thirty-nine years old when he began his trip and fifty-two when the book was published, yet *Travels* has the feel of a young man's book, a young man who has lately slipped the leash of his father's influence and expectations. He writes emotionally about the places he sees, and none of the scenes he witnessed stirred more joy and exuberance in his writing than the pine-covered landscape of northern Florida. On one occasion, he describes his journey in the company of Indian traders from the St. Johns River to the great Indian town of Cuscowilla, near the Alachua Savanna, today known as Payne's Prairie, near Gainesville:

For the first four or five miles we travelled westward, over a perfectly level plain, which appeared before and on each side of us, as a charming green meadow, thinly planted with low spreading Pine trees (P. palustris). The upper stratum of the earth is a fine white crystalline sand, the very upper surface of which being mixed or incorporated with the ashes of burnt vegetables, renders it of sufficient strength or fertility to clothe itself perfectly with a very great variety of grasses, herbage, and remarkably low shrubs. . . . After passing over this extensive, level, hard, wet savanna, we crossed a fine brook or rivulet; the water cool and pleasant; its banks adorned with varieties of trees and shrubs. . . . After leaving the rivulet, we passed over a wet, hard, level glade or down, covered with a fine short grass, with abundance of low saw palmetto, and a few shrubby pine trees [and oaks] . . . : then the path descends to a wet bay-gale; the ground a hard, fine, white sand, covered with black slush, which continues above two miles, when it gently rises the higher sand hills, and directly after passes through a fine grove of young long-leaved pines. The soil seemed here loose, brown, coarse, sandy loam, though fertile. The ascent of the hill, ornamented with a variety and profusion of herbaceous plants and grasses, particularly amaryllis atamasco, clitoria, phlox, ipomea, convolvulus, verbena corymbosa, ruellia, viola, &c.

It's the "variety of the scenery" that excites Bartram's enthusiasm, what he elsewhere characterizes as "grand diversified scenes." Bartram is describing a varied topography that supports several distinct natural communities: a level plain ("hard, wet savanna") "thinly covered" with longleaf; a creek and its floodplain; a poor rolling country ("a glade or down") covered with grass, shrubs, low pines, and scrub oaks; a shrub bog; sandhills; a grove of young pines; and a "magnificent grove" of "stately pines." Diversity delights Bartram.

From Cuscowilla, he traveled to Talahasochte, an Indian town near present-day Tallahassee, again describing in great detail the variety in the landscape. "Now the pine forests opened to view," he writes as he leaves the wet margins of the savanna. "We left the magnificent savanna and its delightful groves, passing through a level, open, airy pine forest, the stately trees scatteringly planted by nature, arising straight and erect from the green carpet, embellished with various grasses and flowering plants; then gradually ascending the sand hills, we soon came into the trading path to Talahasochte, which is generally, excepting a few deviations, the old Spanish highway to St. Mark's." That night the band camped under a grove of live oaks, "on the banks of a beautiful lake," and the next day they traveled over a rocky ridge on either side of which was "the most dreary, solitary, desert waste I had ever beheld." Bare rocks emerged out of white sand, the grass was scattered, and there were only a few trees. Soon he and his fellows "joyfully" entered a region of level pine forests and savannas "which continued for many miles," with ponds of water visible, sparkling through the dark columns of the pines. They ascended again to sand ridges through savannas and open pine forests, negotiating with difficulty through a region dotted with lime-sinks ("cavities or sinks in the surface of the earth"), and camped that night "under some stately Pines, near the verge of a spacious savanna."

The next day the traders descended and continued for miles along a level, flat country over "delightful green savannas" dotted with hammocks of hardwoods. They crossed a wet savanna, a "rapid rivulet," entered more rocky land, and then passed another "extensive savanna, and meadows many miles in circumference" where a herd of Indian horses romped. On one side was a "beautiful sparkling lake." He calls this the best land they had passed through since they left Alachua, featuring a gray, brown, or black loamy soil in the lower portions of the landscape and, on the ridges, "a loose, coarse, reddish sand."

Talahasochte was about ten miles away now. After leaving the "charming savanna and fields," he and his band of traders passed through several miles of "delightful plains and meadows":

> We next entered a vast forest of the most stately Pine trees that can be imagined, planted by nature, at a moderate distance, on a level, grassy plain, enamelled with a variety of flowering shrubs, viz. Viola, Ruella infundibuliforma, Amaryllis atamasco, Mimosa sensitiva, Mimosa intsia and many others new to me. This sublime forest continued five or six miles, when we came to dark groves of Oaks, Magnolias, Red bays, Mulberries, &c. through which proceeding near a mile, we entered open fields, and arrived at the town of Talahasochte, on the banks of the Little St. Juan [Suwannee River].

Travels made Bartram famous, and his idealized descriptions of a lost southeastern Eden have stirred the imaginations of generations of readers. His book influenced the poetry of Samuel Taylor Coleridge and William Wordsworth and the prose of Ralph Waldo Emerson and Henry David Thoreau. He might have been even more famous had he accepted Thomas Jefferson's invitation to botanize on the Lewis and Clark expedition of 1803.

Bartram wasn't the only one who delighted in the beauty of the longleaf forests and savannas. The Englishman Basil Hall, a man of polite society, traveled with his wife from Norfolk, Virginia, to Mobile, Alabama, in the 1820s. The two journeyed in coaches and on foot, and, like so many other travel narratives of the day, Hall's mixed descriptions of the scenery with comments on southern political and social institutions, especially the institution of slavery. His book, *Travels in North America, in the Years 1827 and 1828*, vividly describes the great pine forests he and his wife traveled through, as in this account of a journey from Savannah to Macon, Georgia:

> Our road, on the 22d of March—if road it ought to be called—lay through the heart of the forest, our course being pointed out solely by blazes, or slices, cut as guiding marks on the sides of the trees. It was really like navigating by means of the stars over the trackless ocean. When we had groped our way in this strange fashion for about ten or twelve miles, we came to a place where the slight trace of a road, in the expressive language of the woods, is said to fork, or split into two. . . . Off we went again, over roots and stumps, across creeks and swamps, alternately driving up and down the sides of gentle undulations in the ground, which give the name of a rolling country to immense tracts of land in that quarter of the world. The whole surface of such districts is moulded, by what means I know not, into ridges of sandy soil, gently rounded off, nowhere steep or angular, and never continued in one straight line for any great distance. I have often observed the sea in a calm, after a gale of wind, with a surface somewhat similar, only that in the case of these rolling countries the ridges are not so regular in their direction, and are many times larger than any waves I ever saw. They present no corners or abrupt turns; and though crossed by small valleys, these too have their edges dressed off in like manner, as smoothly as could have been managed by the most formal landscape gardener.
>
> For five hundred miles, at the least, we travelled, in different parts of the South, over a country of this description, almost every where consisting of sand, feebly held together by a short wiry grass, shaded by the endless forest.[4]

His journey through this piney wilderness was "toilsome" and "rugged," he wrote, "but it was a long time before I got quite tired of the scenery of these pine barrens. There was something, I thought, very graceful in the millions upon millions of tall and slender columns, growing up in solitude, not crowded upon one another, but gradually appearing to come closer and closer, till they formed a compact mass, beyond which nothing was to be seen."

The botanical richness of the pine forests of central Mississippi excited Mississippi congressman John F. H. Claiborne on a trip he made in 1841. "Much of it is covered exclusively with the long leaf pine; not broken, but rolling like the waves in the middle of the great ocean," he observed. "The grass grows three feet high. And hill and valley are studded all over with flowers of every hue. The flora of this section of the State and thence down to the sea board is rich beyond description." For another Mississippian, E. W. Hilgard, the open forest was like a park. "The herbaceous vegetation and undergrowth of the Longleaf Pine Region is hardly less characteristic than the timber," he wrote in 1860. "The pine forest is almost destitute of shrubby undergrowth, and during the growing season appears like a park, whose long grass is often very beautifully interspersed with brilliantly tinted flowers."[5]

On his thousand-mile walk to the Gulf of Mexico from Jeffersonville, Indiana, in 1867, naturalist John Muir passed through the pine barrens of Georgia. He described them as "low, level, sandy tracts: the pines wide apart; the sunny spaces between full of beautiful abounding grasses, liatris, long wand-like solidago, saw palmettos, etc., covering the ground in garden style. Here I sauntered in delightful freedom, meeting none of the cat-clawed vines, or shrubs, of the alluvial bottoms." Even the serious-minded forester G. Frederick Schwarz, probing the last of Louisiana's virgin forests in the first decade of the twentieth century, could interrupt his study with a paean "to the picturesque forms of these trees and the charm and beauty of the forest scenes throughout these southern pineries":

> For this is not only a very useful tree, but one of high aesthetic values also. Its long, bushy tufts of needles, the interesting and expressive crown-forms lifted high up on the straight trunks; the openness of the forest, isolating the trees and emphasizing their individual characteristics; the beautiful color contrasts of dark-green foliage, brown trunk and tawny grass; the open glades and occasional wide savannas within the forest, furnishing foregrounds and framing the view; and the magnificent backgrounds of sky and cloud, as seen through the trees or looking across the forest from some low hill—these are only sugges-

tions of the features of beauty and the landscape values to be found in virgin forests of the longleaf pine.[6]

Not everyone loved the pine-dominated landscape of the Southeast. For every traveler impressed by the beauty and diversity of the region, there was another who thought it was monotonous and uninspiring.

One rainy day in Savannah, Georgia, I spent an hour or two in a local historian's antiquarian shop looking over his collection of photographs, stereopticon cards, and historical prints. One print caught my eye. It portrayed the town of Savannah a year after its founding by James Oglethorpe. Based on a sketch attributed to Peter Gordon, one of the original band of 107 settlers, the engraving—*A View of Savana as It Stood on the 20th of March, 1734*—is a powerful scene, a bird's-eye view of a young and aggressive town projecting a tiny finger of Renaissance perspective into the pine-dominated landscape. In the foreground cattle graze on Hutchinsons Island. Sailing ships and skiffs ply their business in the lively harbor while workers hoist bales from the docks forty feet up the bluff. Just on the edge of the plateau, Oglethorpe's tent sits beneath four pine trees, and behind it settlers have erected houses on fenced lots and laid out plats for other dwellings. The outlines of the squares that still distinguish this lovely old town are clearly visible. Two men carry a log into the settlement, perhaps destined for the palisade.

It was the thin cowl of pine forest surrounding the town just outside the palisade that really captured my attention. Oglethorpe had described the site of his new town as being "sheltered from the Western and southern Winds (the worst in this Country) by vast Woods of Pine Trees, many of which are an hundred, and few under seventy feet high," and there they are in Gordon's engraving. It is surely one of the earliest visual representations of the pine forests of the Southeast, and clearly the artist wanted to depict a pine forest—the straight poles of the trunks and the lack of lower branches are clues. "The country all round us is a continued forrest," Peter Gordon wrote in his journal, and in the print the trees crowding into the horizon suggest that immensity, a vast and brooding presence kept at bay by the small settlement's walls.[7]

This impression of an unending, featureless forest was a common theme in accounts written by other members of the Savannah colony. "Take care not to go into the Woods without a Guide," advised one correspondent. "If you ask, how a Country that is covered with Wood, and cut with Rivers and Morasses, is passable," he wrote, "I must acquaint you, that since the Colony was settled, the Ways were marked by Barking of the Trees, to shew where the Roads should go." Despite these aids, it was easy to get lost in the forests near Savannah. Two men who disappeared in the woods were brought to safety after three days by the periodic firing of cannon, a customary procedure. In Purysburg, a settlement on the South Carolina side of the Savannah River, a man who disappeared into the forest was never seen again, and his widow waited a year and a day to remarry.[8]

The miles of sandy soils, the flat topography, and the interminable pine forests, combined with the tedium and difficulty of travel, must have been stupefying for many foreign travelers who had already endured one featureless expanse, the sea, in reaching America. "The country [from Suffolk, Virginia, to Edenton, North Carolina] must be imagined as a continuous, measureless forest, an ocean of trees," wrote German traveler Johann David Schoepf in the 1780s. "A dreary Waste of white barren sand, and melancholy, nodding pines" was how Janet Schaw described her first impressions of land as she traveled up the Cape Fear River in North Carolina after leaving the West Indies. "The road was a deep, wearisome sandy track, stretching wearisomely into the wearisome pine forest," wrote English actress Fanny Kemble of her travels in Georgia in 1838–39. While passing through South Carolina's longleaf forests in 1858, Englishman Charles Mackay wrote:

Where, northward as you go,
The pines forever grow;
Where, southward if you bend
Are pine-trees without end;
Where, if you travel west,
Earth loves the pine-tree best;
Where, eastward if you gaze,
Through long, unvaried ways,
Behind you and before
Are pine-trees evermore.[9]

In 1863, on being asked to summarize her impressions of the country that she encountered on her way to Mobile, Alabama, Lily Langtree, another English actress traveling in America, replied: "I have come through a great deal of low land, flat stretches of red clay country and districts crowded with pine trees. The ride to-day has been peculiarly monotonous. . . . I have never seen anything like this Southern woodland before. It strikes me as immense, entirely too immense, and uncultivated."[10]

If "monotonous" and "dreary" were the most common words in the vocabulary of many travelers in the longleaf pine region, "gloomy" was not far behind. "There was but little to distinguish the one route from the other," Alexander Mackay said, indifferently comparing his trip from Raleigh to Wilmington, North Carolina, to another from Charleston to Columbia, South Carolina. "The whole of this district . . . has, in most places, from the quantity of dark and sombre pitch-pine with which it abounds, [a] gloomy and monotonous aspect." "A dark wilderness of pines . . . [so] dense . . . it seemed as if we had entered a realm of sighing and moaning," wrote a traveler in North Carolina's Cape Fear country in 1853. "It is impossible to resist the feeling of loneliness that creeps over one on entering these silent forests," noted Porte Crayon (David Strother Hunter), "or to repress a sentiment of superstitious dread as you glance through the sombre many-columned aisles, stretching away on every side in interminable perspective."[11]

Today, you have to look hard to find longleaf, but the European conquistadors and adventurers, religious divines and Indian traders, gentlemen and gentlewomen, hunters, naturalists, botanists, and geologists who left written records of their travels in the Southeast were looking *at* longleaf. It was all around them, a land of strikingly different configurations—flat, monotonous stretches of thin sandy pine forests uninterrupted for miles; rolling hills of dense pine forests abutted by lakes, swamps, and other wetlands; forests full of sun and light; shadowy forests. The forests were as featureless and monotonous as the sea or desert, as varied and colorful and delightful as a garden or park, or as sinister as a forest out of legend or European folk tale. If these comparisons of longleaf pine forests seem dissimilar it is because those who beheld them colored them with their own feelings, or because they could not quite get their minds around the immensity of the forest, describing its variety a part at a time like the six blind men feeling the elephant.

TWO

Fire in the Cathedral

Instantly the lightning, as it were, opening a fiery chasm in the black cloud,
darted with inconceivable rapidity on the trunk of a large pine tree, that stood
thirty or forty yards from me, and set it in a blaze. The flame instantly ascended
upwards of ten or twelve feet, and continued flaming about fifteen minutes,
when it was gradually extinguished, by the deluges of rain that fell upon it.
—William Bartram, *Travels* (1791)

You could easily characterize the land of the longleaf pine in terms of the rivers that cross it, or the swamps that give it its unique and mysterious character. From southeastern Virginia to eastern Texas, forty-two major streams cut across the Gulf and Atlantic pinelands on their way to the sea. Wetlands of all types puddle the land. When Hernando de Soto and his conquistadors marched north from Tampa Bay in search of treasure in 1539, they spent a lot of time during the next four years crossing rivers and swamps. In de Soto's day, 35 percent of the southeastern land mass was characterized as a wetland of one type or another. Pure and mixed stands of longleaf pine covered 60 percent of the Gulf and Atlantic Coastal Plains in 1539, but most of the rest was a wetland of one type or another.[1]

Indeed, at one time all of the Southeast had been under water. The Coastal Plain was born of the sea, and the sea still occupies a substantial part of it. Geologists say that the Coastal Plain is in a dynamic tension between its emerged and submerged portions. For the past five thousand years, the emerged portion of the Atlantic Coastal Plain has occupied a wide band ranging from the present shoreline to the Piedmont fall zone; the submerged portion, called the Continental Shelf, has extended from the shoreline to offshore waters about six hundred feet deep. The proportions of submerged and emerged land have changed in response to changing climate, and they are still changing today. At some points during the last 60 million years, Louisiana and Mississippi lay beneath a great inland sea that stretched all the way to Canada. The South Atlantic coastline lay as far inland as Columbia, South Carolina. You could have paddled a sea kayak from Augusta to Macon, Georgia.[2]

During ice ages, when glaciers moved southward across the North American continent and expanding polar ice captured more of the earth's oceans, falling sea level exposed the slope of the Coastal Plain far offshore. At the peak of the last glaciation 18,000 years ago, the South Atlantic coastline lay about 450 feet lower than it does today. As the climate warmed, the seas began their long climb back up the Coastal Plain. This process of sea level rise and fall, triggered by climate change, was primarily responsible for creating the flattened surface of the Coastal Plain. The landscape is so flat that Alexandria, Louisiana, about 125 miles from the Gulf of Mexico, is only 82 feet above sea level. Geologists have found a number of terraces across the Coastal Plain, each the result of a particular epoch of sea level rise, some marked by abrupt scarps that mark the farthest advances of the sea, all of them covered in various layers of sandy sediments deposited by the sea. The coastline today has been in its present position, more or less, for approximately the last five thousand years.[3]

But the rivers have also contributed mightily to the making of the southeastern Coastal Plain. When I drive from Raleigh to Cape Hatteras, North Carolina, or from Hattiesburg to Gulfport, Mississippi, I am traveling over ancient sea bottoms whose sandy sediments were delivered by the region's many rivers. Rivers draining the southern Appalachians far to the west carried the slowly eroding rock to the Atlantic Ocean and the Gulf of Mexico where rising seas could distribute it across the Coastal Plain. Rivers shaped the sandy sediments dropped by the sea, eroding them, over thousands of years, into a complex landscape of level plains, rolling hills, river valleys, and wetlands.

During glacial times, when the coastline lay lower than it does today, rivers could run energetically to the sea, cutting deep channels into the soft sediments. But when the climate warmed, ice caps melted, and the seas crept up the slope, rivers that once rushed over the Coastal Plain slowed down, sluggishly moving in wide looping meanders across one flat terrace after another toward the sea. Rivers that had run swiftly through deep channels across the Continental Shelf were drowned beneath the advancing sea, creating the shallow sounds that characterize eastern North Carolina today. The rivers meandering across the flattened landscape of the Coastal Plain were shallow, and they flooded often across wide alluvial valleys where vast wooded swamps of bald cypress trees and other swamp-loving plants grew, adapted to the wet soils. Water tables rose. The landscape virtually oozed much of the year.

Although the southeastern landscape was born from the sea and carved by rivers, it was fire and storm that effectively shaped the natural communities of the Southeast, especially the longleaf pine forests. William Bartram understood a lot of the varied longleaf ecosystem, yet even he did not suspect the key roles played by these natural processes.

The effects of hurricanes on the forest could be spectacular. In 1528 Spaniard Alvar Núñez Cabeza de Vaca complained that he and his fellow conquistadors, making their way up Florida's western peninsula, encountered "so many fallen trees on the ground that they barred our way." Many of the trees "were split from top to bottom from the lightning that strikes in that part of the world, where there are constantly great storms and tempests." A hurricane that blew out of the Gulf of Mexico in 1772 destroyed the woods north of Mobile for about thirty miles. Up the Pascagoula River "the pines were blown down or broke, and those which had not intirely yielded to this violence, were so twisted, that they might be compared to ropes," according to land surveyor Bernard Romans. J. F. H. Claiborne of Mississippi described a thunderstorm in a longleaf pine forest in a colorful, if perhaps overwrought, style:

> The day was dark and lowering. For weeks nor rain or gentle dews had refreshed the calcined earth. A heavy cloud hung overhead, and its pent-up fury burst upon the forest. The few birds that tenant these silent woods flew screaming to their eyries; some cattle dashed across the hills for shelter. The whole wilderness was in motion. The pines swayed their lofty heads, and the winds shrieked and moaned among the gnarled and aged limbs. A few old ones fell thundering down, casting their broken fragments around; and then the hurricane rushed madly on, tearing up the largest trees, and hurling them like javelins through the air. The sky was covered as with a pall; and lurid flashes, like sepulchral lights, streamed and blazed athwart it. The earthquake voice of nature trembled along the ground, and, ere its running echoes died away, came again, crash after crash thundering forth. But at length, as though weary of the agony, it paused, and the phantom clouds scudded away. The scene around was appalling! Hundreds of trees lay prostrate, while, here and there, others stood shiv-

ered by the bolt of heaven and smoking with its fires. God preserve me from another ride through these giant pines in such a tempest.[4]

The main legacy of lightning storms in the South was fire. Fire posed an evolutionary challenge for plants and animals of the longleaf pine forests: They had to adapt to it or retreat to places where fires burned less often. The fact that longleaf not only survived but also dominated the Atlantic and Gulf Coastal Plains for thousands of years means that it met this challenge successfully. Indeed, fire in longleaf pine forests is like rain in a rain forest. Within the evolutionary pressure cooker of an environment shaped by storm and fire, longleaf developed many extraordinary traits over the millennia, enabling it to dominate an immense area.

By one estimate, 100 lightning bolts are discharged per second over the globe, each one a current that zaps back and forth from cloud to earth dozens of times in less than a second. That adds up to about 8.4 million bolts each day. Within the square mile in which you live, you may expect between 40 and 80 such lightning strikes per year. Where they strike depends on the conductivity of the substance. Some trees are better conductors than others. Oaks and pines are more prone to injury from lightning than beech, even though beech is extremely thin-barked whereas oaks and pines are relatively heavily armored with bark. Bracken ferns near a lightning-struck tree die within a few days of the strike, while another ground-hugging plant such as runner oak seems unaffected.[5]

When Edward V. Komarek Sr. published these findings in the early sixties, he had been the director of Tall Timbers Research Station since 1958 and was on his way to nearly single-handedly promoting the science of fire ecology through the station's annual fire ecology conferences. His main contention, and the subject of many of the papers he wrote, was that fires in longleaf pine were inevitable and even necessary, and that they had left lasting effects on the plant and animal communities of the Southeast.

Lightning storms are common in the South because the semitropical climate is hot and humid and the moisture-laden air masses from the Atlantic and the Gulf of Mexico breed thunderstorms like oranges. Florida has more "thunderstorm days" (defined as a day when thunder is heard) than any other state in the country. Such storms create a lot of fires. Komarek found that one thunderstorm, coming after a drought, produced ninety-nine blazes in Florida alone. Thus a single thunderstorm traveling north out of the Gulf of Mexico could "spew" fires over a vast distance. The rain that accompanies lightning storms doesn't always extinguish fires. After one fire, a log at the Wade Tract Preserve smoldered for nearly ten weeks one summer, despite four or five downpours in the meantime. Such an ignition source could easily start one or several fires.[6]

Fuel for fires is everywhere in longleaf pine forests, particularly blankets of tawny needlefall, large cones, and thick clumps of flammable wiregrass. One of the most visible characteristics of a longleaf pine is its needles, which at eight to fifteen inches in length are longer than those of any other southern pine and often form a graceful corona at the twig ends when viewed from below. They grow in a fascicle or cluster of three from their basal sheath and are shed after the second year of growth. Unlike the leaves of a deciduous tree, longleaf needles clusters are not all the same age. A tree may hold needles that have just begun growing, as well as one-, two- or even three-year-old needles all at the same time. Though there is an almost constant sprinkle of needlefall year round, it is heaviest during the fall and winter.[7]

"Fire can mean a fire in grass, or in leaves, or in herbaceous plant growth, in forest debris, or even in the crowns of trees," Komarek pointed out. "It can travel slowly, quietly, and be as gentle as the whisper of a breeze; or it can travel with tremendous speed. . . . It can be so cool that one can put his hand under it without discomfort, or its heat can be so intense as to nearly consume one."[8]

One nineteenth-century traveler in Georgia, Englishman Basil Hall, left an especially vivid picture of a fire in a longleaf pine forest:

> On the 21st of March, we fairly plunged into the forest, from which we did not again emerge for many a weary day of rugged travelling. The interest of the forest scenery was a good deal heightened by an immense tract of it being on fire. How far this extended we had no means of knowing; but the volumes of smoke filled up the back ground completely, and deepened the general gloom in a very mysterious style. At many places, however, we actually came amongst the blazing trees, and were somewhat incommoded by the heat and smoke. I was amused at one particular spot by seeing a pitch-pine-tree burning in a curious way. The fire had somehow made a hole in the stem, near the ground, and burnt out a passage for itself, of several yards in length, in the heart of the tree; after which, the flame again made its appearance, thus producing a pipe or chimney. There was, consequently, a strong draught, and the poor pine was roaring away like a blast furnace, while its top was waving in the air, a hundred feet above, as green and fresh as if nothing remarkable were going on below.

. . . [There was a portion] of the forest where not only the trees were on fire, but the grass also. It was an exceedingly pretty sight. A bright flaming ring, about a foot in height, and three or four hundred yards in diameter, kept spreading itself in all directions, meeting and enclosing trees, burning up shrubs with great avidity, and leaving within it a ground-work as black as pitch, while everything without was a bright green, interspersed with a few flowers.[9]

This kind of fire was typical of most blazes in longleaf forests, the flames rising perhaps a foot or two above the ground, driven here and there by the wind until finally doused by a heavy downpour or after running up against a creek or wetland. Fires like these could burn for days or weeks and cover an astonishing amount of ground. Fire historian Stephen Pyne reports that in 1898, a fire once burned over 3 million acres in southeastern North Carolina and "barely made it into the back pages of the Raleigh newspapers." The longleaf pine environments were so interconnected that a large fire that ignited in Albany, Georgia, might sweep through Tallahassee, Florida, four weeks later. Today, the longleaf landscapes are fragmented by roads and other artificial firebreaks, and many state forest services are prepared to battle the wildfires that do occur. In Hall's day fires were so common that most longleaf pine forests would have burned every three to eight years. Some burned every two to three years, and others sometimes even burned annually.[10]

The key to the longleaf pine's survival and dominance in such a fire-rich environment was a suite of adaptations—thick bark, large seed size, inconsistent seeding, fall seed sprouting, and slow growth during the tree's early years—that enabled it not only to tolerate fire, but also to thrive with it. All pines are armored by thick bark that helps to insulate the trees' vital cambium from the lethal effects of heat. As a fire licks the bark of a tree, the temperature on the surface can rise to 1,600 degrees Fahrenheit. At a temperature of 140 degrees Fahrenheit, the cambium of a tree is killed. Thus bark can be considered the Maginot Line of a tree's fire defense. It's a natural insulator, with many internal air cells inside a mainly corky material, although its structure and composition differ from species to species. The bark of some tree species is thicker than others, and the thicker the bark, the more fire resistance it provides. Given identical bark thickness, some species more efficiently insulate than others. For example, fire kills the cambium of sweet gum in less than half the time as longleaf pine. As befits a southern pine, loblolly is better insulated than most hardwood trees, but not as well as longleaf.[11]

Longleaf's cones and seeds are also keys to longleaf's success in the region. The tree begins to produce cones when it is about twenty or thirty years old. Longleaf's cone is large, from four to twelve inches in length. Though by no means the biggest pinecone in existence (the sugar pine of California and Oregon produces a cone that can be two feet long), longleaf's cone is the largest of the southern pines. The cones that fall to the ground are the female strobili that hold the seeds. The male strobili are inconspicuous, appearing at the end of the lowest branches and looking very much like catkins. They emit clouds of buoyant pollen that fertilize the receptive female conelets in the spring, with the seeds maturing over a two year-period.[12]

The longleaf's seeds are large, too. They are so heavy that they fall close to the tree, unlike the smaller and lighter seeds of the loblolly pine which fall over a wider territory. This conveys an advantage to an opportunistic species like the loblolly, but the longleaf's large seed has a tremendous survival value, for it is loaded with enough moisture and nutrients to sprout almost immediately on fire-cleared mineral soil and begin its taproot growth before winter.[13]

Within a week or two after falling the longleaf seed pushes out its embryonic leaves called cotyledons. In a good seed year, it is not unusual to find hundreds of seed wings standing straight as soldiers, held aloft by the germinating seed. The autumn timing is not accidental. Lightning-ignited fires occur in any month of the year, but they peak in May and June. For a few months after, the mineral soil will be exposed and the seed, falling from the ripened cones from the end of September to December, will find a hospitable bed for germinating. Thus the autumn seedfall occurs just when the ground is most receptive to the seed and also when a naturally occurring fire is least likely. The seeds need light and room to grow. Without an occasional summer fire that eliminates thick accumulations of grass and debris, most of the seeds would never reach the ground; they would be hung up in the thick tangle of grasses and needles. If they managed to reach the ground, they would be deprived of moisture and growing space.

Other southern pines adapt to fire in other ways. Pond pine and sand pine, for example, also grow in natural communities that burn, but it takes extended dry weather conditions to create a fire hot enough to burn them. The infrequent fires—occurring anywhere from a decade or two to a century—burn intensely and catastrophically and can kill entire stands. Adapting to these kinds of fires, both species evolved a serotinous cone—one that releases its seed only in the heat generated by a fire.

Loblolly, slash, and shortleaf pines have adapted to fire by growing quickly and adding layers of protective bark. By the end of the first growing season, loblolly seedlings may already be seven inches high and will add a foot or more each year thereafter. If they can escape fire for five or six years, only a very hot summer fire will kill them. The frequent fires in longleaf pine communities, however, confined loblolly pines and other tree species to wetter places on the landscape—drains, pond shores, and river floodplains—where fires burned cooler.

Longleaf's strategy was quite different, and it dominated the region precisely because of it. Longleaf begins to grow quickly after germinating, but it puts most of its growth below the ground, not above it. Indeed, before winter the young tree has begun growing a taproot at a sometimes phenomenal rate. In two weeks, it can grow twenty inches long. In ten or eleven months taproots can be eight feet long, and at maturity they may well be much longer. A long taproot is an essential characteristic of a tree species that grows in dry, or xeric, soils, and the capacity to reach the deeper water and reach it quickly gives the longleaf an advantage not shared by many other trees. The long taproot also serves to anchor the tree in the open habitats and sandy soils in which it grows. Perhaps most important for the young tree, the taproot stores food that it will someday need.[14]

For the first few years of its life, a longleaf pine seedling huddles low on the ground and is almost indistinguishable from a grass clump, which is why this early growth is called the "grass" stage. Through its second and third springs it continues to grow slowly. If other southern pines are sprinters, achieving height growth quickly, the longleaf is more like a long-distance runner. It can remain within a foot or two of the surface for two to five or more years, then spurt quickly. In seven years, a longleaf pine may be growing four or five feet in height each year.

Longleaf dominates because of its grass stage. Like other pines, it is quite vulnerable to fire during its first year and even a light, surface burn may kill it. Late in its first year or early in its second it becomes fire-resistant, and now it fully distinguishes itself from the other southern pines. The terminal bud of the grass-stage seedling is surrounded by a bushy sheath of green needles. A fire may burn off almost all of its needles, but incredibly the growing bud is normally safe and the needles quickly grow back. In one experiment, tissue paper placed around seedling buds one to three feet high was not even scorched in a fire, though the needles were burned to within three inches of the bud.[15]

Although a seedling has been known to languish in its grass stage for a dozen or more years, dominant seedlings in most stands start height growth within a few years and grow past breast height in seven years. Drawing on the food stored in their taproots, the pine rockets upward as much as three feet in a single year and perhaps four feet each year thereafter, quickly lifting its growing tip above the level of most fires and gradually armoring its stem with thick plates of bark. After a few years of swift growth, the pine is usually invulnerable to all but the hottest blazes, and it continues to grow at a rate similar to and sometimes even exceeding loblolly and other swifter-starting pines.

Of course, not every seedling follows the same script. Seedlings from the same seed year may well undergo their growth spurt in different years. This unpredictability has frustrated many tree growers, yet a staggered emergence time is another fire adaptation, providing a further guarantee that at least some of its seedlings will survive a hot fire. The longleaf's notorious pokiness is actually critical to its success on the Coastal Plain.

Longleaf produces a few cones and some seed just about every year, but a good seed year occurs just once every five to seven years. This ability to "mast," or vary its seed production over an irregular period, is an adaptation that it shares with other tree species, including other pines and many oaks. In a good mast year, from 85 to 95 percent of the trees will bear cones and half of them will produce more than 50 cones per tree. Each cone releases about 50 or 60 seeds in a good year, although the number of cones and seeds tends to increase with the size of the tree. One ancient longleaf pine twenty-four inches in diameter at breast height and one hundred feet tall produced 150 cones and more than 100,000 seeds. Good mast years are rare and have entered the literature like the great vintage years of a Bordeaux. A big mast year is usually a regional phenomenon, not just a local one.[16]

Tree growers yearn for more regularity in longleaf's seed production, yet these lean years have a practical survival value to the longleaf. Irregular seeding may well be a coevolutionary trait that arose in response to seed predators. Few seeds ever survive to germinate. Even before they are out of the cone, seeds can be destroyed by insects, frost, and squirrels. Once they fall, they are attacked by birds, rodents, large and small mammals, insects, and other predators that can practically wipe out an entire year's production in less than six weeks. Once a longleaf pine's seed germinates, ants may beset the tiny seedling. The seedling's succulent cotyledons—the early needles—

Fire in the Cathedral

[25]

are also palatable to ants and other insects. Young saplings are occasionally beset by bark beetles and southern pine beetles. As much as 99 percent of a seed crop may be destroyed by predators.[17]

Thus if the pine produced a similar number of seeds every year, rising wildlife populations might consume all of the seeds. Not bad for the animals, not so good for the pine, for the survival of the forest depends on at least some of the seeds escaping and germinating. That may be why the longleaf, like many oaks, overwhelms seed predators with a bumper crop one year and starves them for several years following.[18]

The longleaf's conservative seeding strategy is a fine trait for a long-lived conifer in a fire-prone environment. The longer a plant lives, the less important it is to seed often. Since fire is always random, even when frequent, it's bound to burn up a lot of seed and seedlings anyway, so the longleaf wouldn't accomplish much with more frequent seeding.[19]

Other plant species of the longleaf pine ecosystem also have had to adapt to frequent fires, and many of their strategies are similar. Some plants, for example, produce seeds that germinate only after the heat from a fire cracks their hard covers, allowing water to enter. Others seem to have abandoned sexual reproduction entirely. In a fire-prone system such as this, a perennial plant that sprouts vegetatively from a single root or rhizome year after year can quickly take advantage of the nutrients released by a fire. A seed-producing annual, on the other hand, would waste a lot of energy in germinating and developing a root system every year, especially if it were going to be killed off by the next fire. In habitats that are not disturbed by fire as frequently, seed-producing annuals make sense; here perennial plants with fire-proofed underground growing parts seem to have won the day. Fires that consume the top parts of the plants rarely harm their underground parts, and the plants are adapted to resprout quickly. The endangered rough-leaf loosestrife (*Lysimachia asperulifolia*) begins to bloom in early spring, vigorously flowering only where the plant has been burned regularly. It grows from an underground rhizome, producing buds in the fall that become shoots and enable the plant to resprout quickly after fire. The Venus fly-trap sprouts back in as little as two weeks after a fire from its rhizome. Woody shrubs—wax myrtle, gallberry, titi, and saw palmetto—will also sprout after fires. Turkey oaks and bluejack oaks sprout after a winter fire has apparently killed them. Sweet gums and dogwoods will do the same.[20]

The season in which a fire occurs often affects what grows in a longleaf pine forest. Forest managers on the quail-hunting plantations near the Wade Tract traditionally burned their land in February and March each year, after the hunting season ended and before the quail started nesting. For most managers, these annual winter fires are easier to control and they have predictable results: They keep the oaks shrub-sized and the forest open, and for the quail-hunting nimrods, an open forest means better shots during the quail-hunting season. But winter fires also tend to eliminate much of the new pine growth that sprang up after the autumn seedfall, and they don't eliminate the scrub oaks. In time, a regime of regular winter fires may produce a longleaf pine forest with fire-proofed thickets of turkey oaks and other oak species, as well as a sparse understory of grasses and herbs.[21]

Ecologists generally agree that longleaf and its associates evolved with lightning-ignited fires that burned predominantly during the growing season from April to September. Thunderstorms peak in August, but lightning in May and June starts fires that are more successful at killing the scrub oaks. Plant reproduction seems to be tied to fires at this time. Wiregrass will not flower unless it has been burned during the growing season, and growing-season fires trigger abundant flowering and seed and fruit production in a host of other plants. Thus ecologists encourage forest managers to work toward a regime of frequent though not annual growing-season fires.[22]

———

Some ecologists believe that longleaf and its plant associates may well be doing more than defending themselves against fire. They may actually be promoting fires.

This intriguing and controversial theory was proposed in 1970 by a U.S. Forest Service forester named Robert Mutch, and it has entered the botanical literature under his name. Mutch suggested that in certain ecosystems that depended on regular fires, many of the fire-adapted plants might have a vested interest, so to speak, in encouraging blazes to discourage competition. Thus natural selection might favor plants with flammable oils and resins in their leaves and stems. Plant communities with these characteristics would burn readily, discourage other plants with less or no resistance to fire, and improve their own chances for successful reproduction and long-term survival.[23]

As evidence, ecologists point to the highly flammable resin content in longleaf wood and needles. Longleaf is the most resinous of any southern pine, a factor that made it the leading source of turpentine and other naval stores for two hundred years. Longleaf pine needles are also the longest of all the southern pines, and they fall annually.

Another plant that evolved similar pyrogenic qualities is wiregrass, a bunch grass that grows in longleaf pine communities from Virginia to western Alabama where little bluestem grass takes its place. By one reckoning, 90 percent of the understories of longleaf pine communities are made up of grasses, with wiregrass the dominant grass over almost all of the longleaf's eastern range. By looking at a clump of wiregrass, it's easy to see why it burns so easily. Each clump consists of hundreds of individual leaves, most of which are actually dead. Dig down into the tussock and you'll pull up what feels and looks like wood. Indeed, the live leaves have very few green cells; most are woody, fibrous cells, making the entire plant nothing more than a highly flammable piece of kindling.

Is it possible that natural selection favored wiregrass's tinderlike characteristics and the longleaf pine's heavy resin content and annual needlefall, ensuring that the forest burned and their competitors were disadvantaged?

A tidy debate has taken place in the ecological research community over the Mutch Hypothesis. Ecologists of Atlantic and Gulf Coastal Plain ecosystems have clearly been intrigued by the theory, although some of them reject its premise. Because a plant has flammable leaves or oily compounds doesn't mean that they evolved for the primary purpose of burning, these ecologists say. In a lively exchange in *Oikos* in 1984, James R. Snyder of the University of Florida complained that the hypothesis was "Mutch ado about nothing." Many of the characteristics that increase a species' flammability, he wrote, may only be secondary or incidental to other traits that evolved to enhance the plant's fitness in other ways. In the case of wiregrass, for example, the fibrousness of the leaves and its many dead cells may have fit the plant primarily to survive frequent droughts, not just to fuel frequent fires. So shallow-rooted is the plant, and so droughty are the upper layers of the soil most of the year, especially in sandhill communities, that one might wonder how the plant survived at all. This was the problem that intrigued B. W. Wells in the 1930s, and he explained its peculiar characteristics as adaptations to drought. "Thus in wire-grass we see a striking illustration of how plants meet low-water conditions: they reduce the relative amount of living cells, and besides check as best they may the evaporation of water."[24]

It's a chicken-and-egg debate. Though it is difficult to identify the evolutionary origins of the flammable plant materials in longleaf pine communities, it's unmistakably true that the ecosystem is self-reinforcing because of these flammable materials. With frequent growing-season fires, longleaf and its pyrogenic associates will survive and produce fuels for even more fires, thus maintaining an open environment favorable to themselves and unfavorable for fire-intolerant competitors.[25]

Not surprisingly, the distinctive wildlife species of the longleaf pine grass-land system are as dependent as plants on an open landscape created and maintained by frequent fire, and they are as vulnerable as plants to its absence. Fires apparently do not cause any diminishment in the number of species in a grassland habitat. Grassland animals have evolved with fires, and they know how to deal with and benefit from fires.[26]

Large mammals and most birds seem to have the most effective means of escaping fire — they move away from it, or, since fires burn in patches most of the time, they flee to an unburned green oasis left by the patchy fire. Nestling birds are especially vulnerable to a fire, particularly since many bird species common to longleaf pine habitats nest on the ground or in low shrubs. Growing-season fires will kill some young birds, yet in the long run, ecologists say, it doesn't matter because fire improves the habitat. Fire eliminates the stalks of grasses and other plants that hamper birds' efforts to scratch for their food. A seed eater like bobwhite quail doesn't do well in heavy plant growth. Insects are more abundant after spring grassland fires and less abundant in dense thickets that have grown up in the absence of fires. Many ground-nesting bird species will renest if their nests are destroyed by fire. Bobwhite quail will sometimes renest three times a year.

Wild turkeys generally won't renest, producing a single clutch of eggs that hatch in mid-May. A growing-season fire may well destroy the nest and eggs, but turkeys respond positively to the open forest produced by growing-season fires. Some biologists contend that spring fires stimulate seed-producing plants that turkeys, quail, gopher tortoises, and many other species favor; increase insects that some animals eat; and increase the protein content in plants browsed by deer.

Some wildlife species are even attracted to fire rather than flee it. One afternoon, while witnessing a prescribed fire in North Carolina's Sandhills Game Land, I noticed a red-tailed hawk circling overhead. While two technicians ignited the wiregrass, another lightly disked a grassy road to provide a firebreak between burn compartments. Even as the first wisps of smoke curled up into the air, the hawk swooped into a pine tree overhead and peered at the work going on below him. "That's ol' Red," one of the technicians told me. "He always comes around when we burn." The hawk was instinctively drawn to the fire to find rodents moving out of harm's way. Instead of fleeing the flames in fear, the hawk saw an opportunity for an easy meal.

Even the hawk's prey are not helpless in the face of a typical fire. Mice, rats, and voles — part of a primary food chain in a grassland ecosystem like the longleaf pine — follow a variety of strategies to avoid fire, depending on the circumstances. Rodents are burrow excavators and when fires occur, they can duck into them and escape the flames and heat from the typical grass- and needle-fed fire. If they are caught in the open, however, they can hunker down in a low spot and let the fast-moving fire run right over them. If they are lucky and the burning conditions are right, they might suffer nothing more serious than singed fur.

Edward Komarek of Tall Timbers believed that cotton rats could sense when fire approached and the direction from which it came. Given the burning characteristics of most longleaf pine forest fires, these animals had enough time to place their young in little "popholes" where they would be safe. He also found pathways from the unburned islands to the burned areas where the cotton rats were feeding on the tender grasses that began to emerge from the blackened soil just days after the fire.

Another bird species not generally considered a resident of longleaf pine forests foraged in dead pines killed by fires and lightning. The extinct ivory-billed woodpecker is best known as a native of the bottomland hardwood swamps that grew about the myriad rivers of the Southeast. Yet in some places, these large woodpeckers nested in swamps and foraged in the surrounding pine forests for wood-boring insects, their favorite food. Some of the original, old-growth pine forests had high populations of wood-boring beetles that attacked trees that had been injured or killed by lightning or by lightning-ignited fires. Such an abundant source of food didn't go unnoticed by the ivorybills. Because of their great size, ivorybills required a huge territory of as much as six square miles per pair. That amount of forest could support 36 pairs of pileated woodpeckers and 126 pairs of red-bellied woodpeckers. What hurt the ivorybills was a trait they had developed during their evolution. They singled out large, old, and recently killed trees and scaled away plates of bark to reach the first layer of grubs and beetle larvae underneath. The ivorybill's feeding strategy worked fine with a huge forest that contained a lot of old fire-killed trees, but as these forests disappeared, the ivorybill lost an important food base.[27]

A longleaf pine forest may seem parklike, yet it is really a combat zone in which longleaf vies for survival with rival pine species and the resident oaks. As long as fires burn regularly and keep the ground open, there's no contest. Longleaf and its fire-dependent associates dominate the landscape without serious competition. In the absence of fire, however, competitive species creep up from the wet places where they have been confined. With its prolific, annual, and predictable seed, loblolly quickly seeds in the empty places; shrubs and hardwoods, formerly controlled by fires, grow taller and more numerous, with thicker stems. There's less fine fuel on the ground to burn, and oak leaves burn less readily and more coolly than pine needles. The oaks can fireproof themselves and create conditions for their own dominance.

As hardwoods muscle into the open spaces between the pines and up into the canopy, they outcompete the pines for root space, and the grasses, herbs, and legumes are eliminated along with their provender of seeds and berries. These deciduous forest conditions discourage the populations of gopher tortoises, Bachman's sparrows, red-cockaded woodpeckers, Florida scrub jays, and fox squirrels that could once make a living here. Now blue jays, red-eyed vireos, hooded warblers, flying squirrels, and a suite of species that favor shrubbier hardwood forests find a home.

The longleaf, with its unpredictable seed years and its need for bare soil in which to germinate, cannot regenerate. Veteran longleaf pine trees are still there, but the little seedlings can't grow and the old pines die without successors. The grasses, shrubs, and wildflowers, with their need for sunlight and root space, are shaded out and cannot reproduce. The longleaf pine community may deteriorate so much that it may not be capable of being restored. What was dominant for five thousand years because of fire essentially disappears without it.

THREE

A Wondrous Diversity

In recent years a new emphasis has appeared in the field of plant study,
which involves the attempt to understand the plant in relation to its environment.
It tries to answer such an important question as why plants grow where they do and
the equally significant one of why they are not present when absent from an area.
Organism and environment constitute the real whole, so that ecology, the science
which deals with both in their relation to each other, is becoming increasingly
valuable as a major science helping us to understand the world about us.
—B. W. Wells, *The Natural Gardens of North Carolina* (1932)

"Xeric sandhill scrub community," says Richard LeBlond, gesturing to the mounds of whitish sand, scattered longleaf pines, squat turkey oaks, sparse wiregrass, and dark patches of lichen. This sandhills site is a harsh environment. Save for the stunted trees, we could easily be striding along the crest of a beach dune. It's surprising anything is growing here at all.

The late September day is unusually warm. When I left Raleigh this morning, I dressed for a chill that has melted away. LeBlond, an ecologist with the North Carolina Natural Heritage Program, is lightly dressed in a white T-shirt, and with deer season already in full swing he wears a blaze-orange baseball cap.

We are in Brunswick County, in the extreme southeastern part of North Carolina. Brunswick County is bordered on the east and north by the Cape Fear River, on the south by the Atlantic Ocean, and on the southwest by the Waccamaw River. The vivid contrast between this sandy site with its dwarfed trees and the beautiful rolling hills of the Wade Tract Preserve with its lofty old pines has already taught me an important lesson about longleaf pine forests: They don't all look alike.

For many people in the longleaf region, the terms "flatwoods," "sandhills," and "savannas" soon became more or less standard ways of differentiating among three broad types of longleaf. Plant scientists have used increasingly sophisticated criteria to differentiate among these landscape types, identifying themselves in the process as either "lumpers" or "splitters." The lumpers tend to overlook the slight variations between similar landscape types, whereas the splitters are hesitant to lose the differences. Some operate as both lumpers and splitters at different times and for different audiences. B. W. Wells, the pioneer North Carolina ecologist writing for a popular audience in 1932, reduced the state's entire vegetation, "approached from the ecological point of view," to ten basic plant communities, including two longleaf pine types. Yet when writing for scientists, Wells had been a "splitter," subdividing the sandhills vegetation itself into five different classifications based on the amount of moisture present in the soil.[1]

Later ecologists have tended to be splitters, making finer and finer discriminations. In 1990 Michael Schafale and Alan Weakley, ecologists with the North Carolina Natural Heritage Program, identified 8 longleaf pine communities in the state. Later work by the Carolina Vegetation Survey has come up with about 40 distinct longleaf community types in the Carolinas, and survey leader and University of North Carolina ecologist Robert Peet estimates that as many as 150 types exist in the Southeast as a whole.[2]

All of these communities contain longleaf pine and a grassy and herb-rich understory. They are all fire-dependent, as well, but each differs markedly in its plant composition. The differences depend on a number of factors — moisture levels, the soil's mineral or organic content, and how often and how intensely fire burns through. Thus turkey oaks may be common in one community and all but absent elsewhere. Longleaf pine trees might well grow tall and robust in one setting, but short and squat in another.

Longleaf pine's ability to grow in a wide variety of settings is the key to understanding the astonishing biological diversity of the ecosystem. Indeed, what William Bartram sensed in his five-year ramble through the southeastern woodlands was a botanical richness and variety that ecologists have only recently begun to come to grips with. Though many travelers justly described as monotonous the *experience* of traveling through the vast pinelands of longleaf, the landscape itself and its plant life were anything but monotonous. Every subtle change in topography, in moisture level, and in soil type brought forth scores of different herbs, shrubs, and grasses and yet one tree species common to all these communities — longleaf pine.

Yet not even Bartram could have surmised what ecologists are daily documenting — that longleaf is one of the most biologically diverse ecosystems on earth.

On a surprisingly short walk, you can easily pass through three or four differ-ent longleaf pine communities. You might even notice them if you were alert to the changes in vegetation. This is what LeBlond is going to demonstrate for me as we traverse a gradient from a sandy ridgetop to a wet pocosin a few hundred yards away. "This is one of the sandhills communities," he says as he strides across the highest point of our gradient, scattering fallen turkey oak leaves as he goes. "These turkey oaks are the indicator species for the xeric sandhill scrub community. This is about as dry and droughty a spot as you can get in the longleaf pine region."

Imagine a place of deep coarse sand, in some places so deep that it is often sheer labor to walk through. Imagine a place, moreover, that gets as much rain as Seattle every year yet has characteristics similar to a desert, for water passes quickly through the coarse sand, leaching out what few nutrients exist in the soil.

Such conditions have challenged plant and animal life to their limits. So dry (xeric) is this community that only two tree species have met the evolu-tionary challenge, the longleaf pine and the turkey oak (*Quercus laevis*). Many of the pines have flat tops and are barely forty or fifty feet tall. Most of the turkey oaks are dwarfed and deformed ("scrub"), although they have grown to tree stature in some areas. Wiregrass barely makes an appearance. Some shrubs—the dwarf huckleberry (*Gaylussacia dumosa*), staggerbush (*Lyonia mariana*), and small blueberry (*Vaccinium tenellum*)—are able to live here, along with a few hardy plants with deeply penetrating taproots such as Caro-lina ipecac (*Euphorbia ipecacuanhhae*).

B. W. Wells called the North Carolina Sandhills "Deserts in the Rain," amazed at how anything could survive where the soil was bone dry an hour after it rained, where it was fried in summer by the sun and subject to fires, where nutrients were few and water was nearly nonexistent, and where plants had to work hard to get both. The only plants that could survive the frequent droughts were those that reduced their water needs or their water loss. Conditions such as these eliminated most of the broad-leaved trees, which transpire freely, but are perfect for pines. The longleaf pine itself, like other pines worldwide, is well adapted to drought. Narrow needles and a waxy needle covering reduce water loss, and its deep taproot finds what-ever water there is below the surface. Physiologically, it's a hardier tree in dry environments than all other leafy trees except the turkey oak. This oak solves the challenges of living in sandy soils in a number of ways. The deep lobes of its leaves and their thick, waxy coating reduce transpiration, and it holds its leaves in a vertical position, rather than outstretched like other broad-leaved trees. Wells explained that the vertical position was the turkey oak's adaptation to the lethal intensity of light reflected from the sandy surface at midday. By experimentally forcing turkey oak leaves into a horizontal position, he discovered that the leaves' tissue soon burned.[3]

On a little knoll I notice a prickly pear growing. Its thick, fleshy leaves, like those of all cactus plants, store water, another adaptation to the droughty ridges. Succulence is only one of a variety of adaptations to drought among the plants growing on these dry soils. The Carolina ipecac forms thick roots that store water. October-flower (*Polygonella polygama*) drops its leaves in dry, hot weather. Ironweed (*Vernonia angustifolia*) has a narrower leaf than most of its cousins. Wiregrass, like longleaf, concentrates most of its water-absorbing root growth in the first foot of soil, to make the most of whatever moisture is available. With its few green and growing leaves, wiregrass requires little water to survive, although on these ridges it seems less vigorous than on wetter sites.[4]

These xeric traits are well known among desert plants worldwide, indicating how desertlike these sandhills ridgetop habitats are. But as LeBlond and I walk down from the highest ridge, a couple of new oak species appear. The narrow leaves of one squat tree resemble those of a stunted willow oak.

"Bluejack oak," he says. It's the indicator species for another community—the pine scrub oak sandhill community. "Turkey oak dominates the xeric sandhill scrub community, the driest sandhill community, whereas bluejack inhabits a site not as dry. The bluejack community has a greater species diversity than the xeric sandhills, but not as great as the wet pine flatwoods or the savanna."

It's a subtle change. Longleaf pine is still here; so is wiregrass, which I notice is growing a bit more densely with fewer open patches of sand visible between the clumps.

The soil is more moist (mesic). The sandy soil may well be finer or mixed with loam. Either condition will result in the soil holding more moisture for a longer period. Or possibly there is a clay layer beneath the surface that has perched the water table higher and encouraged plants that are adapted to slightly wetter soil. In quick succession blackjack oaks (*Q. marilandica*) and dwarf post oaks (*Q. stellata*) appear among the bluejack oaks (*Q. incana*).

"Bluejack oaks and blackjack oaks like loamy soils," LeBlond tells me. "Dwarf post oak likes better soils than post oak."

There's a little more water for plants here than at the top of the ridge, and the soil is more loamy, changes that have made a great difference in the kinds of plants that can grow here. You can find persimmon trees now and even a sweet gum, species that would never have survived in the xeric conditions of the sand ridge. A little shrub such as dwarf huckleberry, with its thickened taproot that acts as a water storage organ, is common in this community. Legumes appear.

We walk farther downslope. "We're in the wet pine flatwoods community now," says LeBlond. "You don't see as many oaks here and the wiregrass is a lot thicker." There are more shrubs. The landscape is flattening out, bordered on one side by a thick palisade of evergreen pocosin vegetation.

We push through wiregrass interspersed with low shrubs, mostly red bay shrubs (*Persea borboni*) with their grayish-green berries. A group of young longleaf pines are growing in a water-filled depression next to red maple seedlings.

"Flatwoods" is a term that lacks a precise definition. Most biologists agree that flatwoods are open woodlands dominated by longleaf pine where the water table hugs the surface, making the soils more moist. They are more common along the flatter outer Coastal Plain of North Carolina than in the rolling hills of the Fall-line Sandhills. In Florida, pine flatwoods cover about half the land surface, occupying areas with very little relief. Slash pine and pond pine often join longleaf in Florida's flatwoods, although not in North Carolina's. (Slash pine doesn't occur naturally in North Carolina.) Flatwoods are typically shrubby, although they may have been less so in the days when fires swept through them more frequently. Georgia and Florida flatwoods contain saw palmetto (*Serenoa repens*), gallberry (*Ilex coriacea*), fetterbush (*Lyonia lucida*), dwarf huckleberry, wax myrtle (*Myrica cerifera*), and dwarf live oak (*Quercus minima*). North Carolina's pine flatwoods are thick with wax myrtle, gallberry, and sweet bay (*Magnolia virginiana*) shrubs growing under the pines.

Flatwoods can be wet or dry, and they burn more often than sandhills. Wax myrtles, gallberries, and saw palmettos, all possessing waxy, evergreen leaves, are especially flammable, making fires here more intense than in savannas or sandhills. Sandhills and flatwoods communities can be distinguished by what occurs in the absence of fire: flatwoods turn shrubby, sandhills thicken up with scrub oak trees.

We walk the edge of the flatwoods, adjacent to the pocosin, which is a nonlongleaf community densely thicketed with evergreen shrubs over a peaty soil. The soil is saturated here, and in some spots it's wet up to our ankles. These wet ecotones between the forest and the wetland types that border them—pocosins, titi (pronounced tie-tie) drains, wet swales, ponds

—are common in the longleaf pine region, and it's here that biologists' pulses quicken.

"Most of the rare plant species are found in the wetter longleaf pine systems—the wet pine flatwoods and the savannas," says my guide. "And right here in the ecotone is where these rarities are most concentrated." Because they host so many rare species, the ecotones are matters of deep concern to ecologists. They are the frontiers of the longleaf pine forest, where shrubby invasions from the wetlands take place in the absence of fire and where the greatest toll in rare species is measured.

"In the absence of fire, these are the areas that are first to lose their integrity," notes LeBlond. "The shrubs move upslope out of the pocosin and become dominant. That will crowd out the rare species." The wet ecotones are also where land managers sometimes create firebreaks by scraping a corridor free of vegetation. Managers are learning that such firebreaks are often unnecessary, that the wetness of the gradient itself is normally more than enough to dampen most fires.

As we walk back to our cars I'm realizing that the elevation change from the top of the ridge, where we had begun our walk this morning, to the bottom, where we ended it, was no more than a few feet. Yet we passed through several natural communities, each distinguished by slightly different soil types and soil moisture. Fires run through each one on a regular basis, but this too may differ in each area. On top, amid the deepest sands, the sparse longleaf and wiregrass provide less fuel for fires and thus they burn less frequently and less intensely. Where the soil gathers moisture downslope, longleaf and wiregrass grow more densely and they fuel bigger and more frequent fires. These conditions encourage more plant species. As soil, moisture, and fire intensity change, plant composition does too, so that each of these communities supports a distinctly different suite of plants. The differences between the dry, sandy area on top of the slope (xeric sandhill scrub), the moister areas in midslope (pine scrub oak sandhill), and the wet pine flatwoods at the bottom seem subtle, perhaps, but each is a different longleaf pine community.

━━━━━━━

Looking at the species richness of communities along a gradient is one way of measuring the biodiversity of the longleaf pine ecosystem. Another way is to enlarge the scale to include, say, the entire Southeast. This measure of diversity would embrace the deep and desertlike sandy soils of the Fall-line Sandhills, the rich and loamy soils around Thomasville, Georgia, the rolling hills of southern Mississippi, and the high elevations of the Cheaha Mountains in Alabama. Longleaf pine is a continuous feature of these very different areas, and the number of species taken in totality expands enormously with the geographic change.

The most common way of measuring the biodiversity of a landscape, however, is to count the number of species within a particular community. This endeavor has been done most spectacularly in the savanna communities.

In 1920 B. W. Wells, traveling the Atlantic Coast Line Railroad from Raleigh to Wilmington, looked out his car window near the town of Burgaw and gazed at a wide savanna of about 1,500 acres. He was awed by the savanna's size, but its wealth of blooming wildflowers astonished him. In the years following his railroad car epiphany he visited and revisited the Big Savannah, as he called it, many times. With his colleague I. V. Shunk, Wells documented the nearly yearlong cavalcade of blooming plants that occurred in the Big Savannah, from the bog dandelions and sun-bonnets in February and March to the pogonias, grass pinks, and other orchids of May; the lobelias, meadow beauties, and milkweeds of June; the blazing stars and tofieldias of September; and the asters of November and December. Wells identified the large number of species that comprised the floral feast outside his train window, and he studied the ecological factors — water table variations, soil nutrients — that contributed to it.[5]

Later ecologists studied savannas with an eye to understanding how the richness of these communities compared to other biodiverse systems. Tropical rain forests, for example, are considered the most biologically diverse ecosystems on earth. Three hundred tree species have been found in a hectare (two and a half acres) of Peruvian rain forest — there are 700 tree species in all of North America. If longleaf pine diversity was measured by tree species, of course, it would be considered biologically poor, given the sheer dominance of the longleaf and the absence of other tree species. In the 200-acre Wade Tract, 97 percent of the trees are longleaf pines, but there are nearly 400 plants in the ground cover. The species richness of the longleaf pine ecosystem lies in the ground cover. You have to get on your hands and knees to discover the richness of this ecosystem.[6]

In the early 1980s Joan Walker, then a graduate student at the University of North Carolina working with Robert Peet, began to study the species diversity in a number of longleaf pine savannas in the Green Swamp, a 15,552-acre preserve in southeastern North Carolina owned by The Nature Conservancy. She counted the species at scales ranging from the very small to the large. In her one-square-meter plots, she found an average of thirty-five species, with a high of more than forty. Longleaf pine seedlings grew next to numerous grass species and joined orchids, lilies, asters, carnivorous plants, and sedges.

These counts were off the charts, "a level of small-scale species diversity higher than any previously reported for North America and roughly equivalent to the highest values reported in the world literature," as she and Peet announced in 1983. "In counting thousands of square meters elsewhere I had never found anything over seventeen," recalled Peet one morning in his Chapel Hill office. Even the tall-grass prairies, as rich as they are, never produced more than an average of eighteen species per square meter, he said. "Obviously, something unusual was going on."[7]

How could so many species coexist in the same small area? Normally, as plants compete for light, the faster-growing plants become dominant and the slower-growing ones die out. In the savannas, dozens of plants flourished side by side. Walker's study sent Peet looking at other places where "high species packing" had been recorded, and he found that most of them were on infertile grasslands in Europe. In each case, something was taking the tops off the plants. In Europe, it was hay mowing and cattle grazing; in the longleaf pine savannas of the southeastern United States, it was fire.

"On these savannas, chronic fires reduce the intensity of competitive interaction, particularly for light," Peet said, "so that a species isn't constrained only to that piece of the gradient they grow best on, but instead is more broadly distributed. If you're competing for light, the plants that get to the top get all the light, and the ones underneath don't get any and they lose. On the other hand, if you're competing for soil resources, it doesn't matter whether you're big or little; it's how much root-surface area you have going after a nutrient like phosphorus. And little plants have just as good a chance at making it as big plants."

Quickly sketching a couple of curves like camel humps on a sheet of paper, Peet explained that if you wanted to graph how the number of plant species relates to soil fertility, you'd find that in grasslands they take on a bell-shaped curve. The more fertile the system is, the more species grow until a certain fertility level is reached—say, pastures that are heavily fertilized to make them grow more grass. At this point, the number of plant species begins to decline because the little plants are being shaded out by the taller grasses. But in longleaf savannas, soils are intrinsically infertile and fire reduces the amount of aboveground biomass, enabling more small plants to flourish.

The results are some of the most extraordinary plant diversity on earth. Peet's later studies showed that at small scales, longleaf pine communities in both North Carolina and the Gulf Coastal Plain rivaled and even exceeded the diversity of tropical rain forests. He found 12 species in a half-dollar-sized site in Mississippi. As the scale grew larger, between 100 and 1,000 square meters, rain forest diversity began to pull away somewhat: at 100 square meters there were 95 plant species on a North Carolina savanna and 123 in an Equadoran rain forest (236 in a Costa Rican rain forest!). At 1,000 square meters, the Equadoran rain forest registered a gaudy 356 species while a longleaf pine woodland in the Gulf Coastal Plain showed a modestly spectacular 140. (Other ecologists, counting species in a 1,000-square-meter longleaf site in Louisiana, have found 175 species.)[8]

But it is not just the number of species in longleaf pine communities that has impressed ecologists; it's their rarity and their limited distribution. Students of Coastal Plain ecology are struck by the sheer number of endemics in this region. An endemic is a plant or an animal that is tied exclusively to a particular area large or small, or to a particular ecosystem, and doesn't occur elsewhere. Sometimes endemics require very special ecological conditions and can thrive nowhere else. Some species are endemic because they have only recently evolved and haven't begun to migrate, some are part of an older population that migrated and then was cut off by some natural barrier. Others may be very old and reflect ancient distributions. Many of the savanna plants occur nowhere else but in longleaf pine communities; often they are quite local in their geographic distribution.

I questioned Alan Weakley at length about longleaf pine endemics and what they suggest about the ecosystem. He noted, first, how incongruous it is that one would even consider the longleaf region to be diverse. Compared to a rich floristic state such as California, whose natural communities include the high peaks of the Sierra Nevada, the Central Valley, the fog-shrouded coast, and the Mojave Desert, the Southeastern Coastal Plain has a fairly uniform topography. "You've got longleaf pine stretching for 1,500 miles from southeastern Virginia to eastern Texas," Weakley said. "At a quick glance, one longleaf pine savanna looks a lot like another longleaf pine savanna."

Yet what's unusual about the Southeast is that most of the plants are confined to the region, and many of them are endemic to the longleaf pine ecosystem. Weakley and botanist Bruce Sorrie have estimated that about 1,630 plants are restricted to the Southeast, and more than half of them (891) are endemic to the longleaf pine ecosystem.[9]

Why does the Southeastern Coastal Plain, and the longleaf pine ecosystem in particular, a region that has little relief and that looks so monotonous and uniform, rival only California in diversity and endemism? "The richness of the Southeast coastal plain has not been adequately explained," Weakley said. "Part of the answer has to do with a trend toward more diversity in vascular plants as you move south to the Equator. The Southeast would thus be naturally rich."

Another factor in the Southeast's diversity can be explained in biogeographic terms. During the Pleistocene, ice descended periodically from the polar region and moved like a giant road grader over much of the northern part of the continent, burying everything in its path that couldn't move and driving everything south that could. The glaciers extended south to a line ranging from Long Island through Pennsylvania, the Ohio and Missouri Rivers through North Dakota, northern Montana, Idaho, and Washington. North of this line lay a bleak terrain of mile-high ice. South of it, the climate changed profoundly and so did the composition of the forests—just five thousand years ago boreal forests of hemlock, jack pine, and spruce, more common today in northern Michigan and Canada, grew over parts of the Southeastern Coastal Plain. Over the course of four advances of the glacial ice flow, the Southeast became a refuge for many species, some of which persisted when the glaciers retreated and helped to revegetate the North during the interglacial stages. Because the southeastern plant communities were spared the homogenizing effects of glaciation, they enjoyed long-term stability over thousands of years. During that time, plants could evolve ever more refined adaptations to compete in the unique microhabitats of the Southeast—adaptations to the frequent fires in the region, for example; adaptations to soils that were saturated much of the year, or that were as dry as deserts, or that lacked essential nutrients. The more time these plants had, the more affinities they could develop to local conditions and the more endemics there were.

It's one reason why so many grasses, though not at all related, converged on a similar morphology. "Grasses like *Aristida* (wiregrass), or *Sporobolus* or *Muhlenbergia* all have relatively narrow blades and hardened leaf bases that serve to fireproof the growing tip of the rhizome," Weakley said. "These are obvious fire adaptations. When you have that kind of situation it's telling you that there are ecological forces that are favoring that morphology and driving different species toward it."

What's more remarkable to Weakley than the sheer number of endemics is the patterns of endemism within longleaf communities. "Those thousand plant species are not *scattered* over this region," he told me. For reasons no one quite understands, there are certain hot spots for endemics, places where endemics seem to concentrate: the region comprising southeastern North Carolina and adjacent South Carolina, for example; the Florida peninsula; the Florida Panhandle extending into southwestern Georgia; and western Louisiana and adjacent eastern Texas.[10]

"Why do you have a suite of thirty or forty species that are endemic to southeastern North Carolina and one or two counties in adjacent South Carolina?" Weakley asked. "Why can't you find those species, like the Venus fly-trap, all over the Southeast?"

The Venus fly-trap (*Dionaea muscipula*) is a unique plant with amazing adaptations to its environment. Charles Darwin called it "one of the most wonderful [plants] in the world." Growing from a bulb below the ground surface, the plant consists of a number of narrow leaves that may grow about four inches long. Each of these leaves broadens at the tip into two lobes that join at nearly a ninety-degree angle. These are the traps. Fringing each of the lobes are anywhere from fourteen to twenty-one teeth that interlock when the trap snaps shut. On the inside of each broadened lobe are three bristles, the trigger hairs. A fly crawling around the interior of the trap will inevitably bump into one of the trigger hairs. A single bump does nothing. But if the hair is struck a second time within about forty seconds, the trap snaps quickly imprisoning the fly. The struggles of the insect inside the cage cause the two halves of the leaf to increase pressure on each other, forming a digestive cavity. Over the next day or so, enzymes help digest the soft parts of the animal and in a week or more the trap opens again, revealing the insect's indigestible hard parts.

Wonderful but rare, for the Venus fly-trap can only be found within a 100-mile radius around Wilmington, North Carolina. Why? Ecologists don't know.

Another Carolina endemic is the rough-leaf loosestrife, which appears only in the Coastal Plain and Sandhills of southeastern North Carolina and northern South Carolina, and has been federally listed as endangered in both states. Historically, it has never been found anywhere else, although populations were probably more contiguous at one time when it was burned more frequently. Today the few population centers that remain are functioning as islands in a discontinuous habitat, and so there is some concern about their genetic viability over time. The Sandhills pyxie moss (*Pyxidanthera barbulata* var. *brevifolia*), a mat-forming evergreen shrub with mosslike leaves and white or pink flowers, has been seen only in a few Sandhills longleaf pine locations in North Carolina and South Carolina, and most of its populations are limited to North Carolina's Fort Bragg, a military installation, where it seems to thrive amid fires caused by exploding ordnance.

There are other interesting plants with a restricted distribution. The flameflower (*Macranthera flammea*) is a spectacular plant found in Gulf Coastal Plain bogs from Panhandle Florida to southeastern Louisiana. It belongs to the snapdragon family and grows nearly ten feet tall, with large orange blossoms pollinated by hummingbirds.

"There are also good examples of endemics in the legume family," Weakley said. Legumes are high-protein peas that feed many wildlife species. "Baptisias, for example, comprise some very narrow endemics. *Baptisia arachnifera* only occurs in two counties in Georgia. *Baptisia cinerea* is endemic to the Carolinas, although it does occur in one county in Virginia and one in Georgia." He rattled off a half dozen other Baptisias with limited distributions in the longleaf pine region.

"A plant genus like *Carphephorus*," he continued, "is a genus that is entirely endemic to longleaf pine habitats, and each species is pretty narrowly distributed. The species *bellidifolius* is essentially found in the Carolinas, one county in Virginia and Georgia; *carnosus* in Florida only; *corymbosus* in south Georgia and Florida; *paniculatus* is Atlantic Coastal Plain, North Carolina to Florida and Alabama; *pseudoliatris* is a Gulf Coastal Plain endemic, from Florida and southwest Georgia to western Louisiana; and *tomentosus* is another Atlantic Coastal Plain thing, Virginia to Georgia."

The reason for such narrow distributions is that all of these species probably descended from a single ancestor and evolved over an immense amount of time when different populations in local habitats began to diverge into new species. Once a single species growing over the entire extent of longleaf pine, for example, the *Carphephorus* evolved in different directions to take advantage of local conditions. *Carphephorus bellidifolius* likes the dry, sandy habitats of the Carolina Sandhills; *pseudoliatris* prefers wet habitats in Gulf Coast pitcher plant bogs.

"In looking at endemics what we're specifically looking at is the biological uniqueness of the region, more than its richness," Weakley noted. "Endemism tends to be an indicator of age and evolutionary history; it's a property that has developed over time. It's partly a reflection of unique habitats, which cause endemics to develop; partly a result of new endemics which have recently evolved; and partly a result of paleoendemics that evolved a long time ago and have hung on."

―――――――

Because so many plants are so narrowly distributed within the longleaf pine ecosystem, many of them are considered rare and endangered. The North Carolina Natural Heritage Program has counted 107 species of rare plants and 13 rare animal species occurring in these habitats. About 600 plant species — or nearly a quarter of the state's total — have been found in longleaf pine communities. In the Southeast as a whole, one study found 389 rare plant species associated with longleaf pine, almost half of which were vulnerable to extinction.[11]

Species diversity is threatened most, biologists say, by a number of regional forces that include drainage of wetlands, habitat destruction (for housing or commercial development, for loblolly pine plantations), as well as a lack of frequent fire that is necessary to maintain the species, especially species that are critical for wildlife.

Longleaf pine *community* diversity has suffered, too. Florida's vast peninsular sandhill communities once grew millions of longleaf pine trees, but millions of acres have been converted to housing developments, citrus groves, and pastures. Most mesic longleaf pine communities throughout the Southeast have been converted to farmland or other uses over the centuries since settlement. Longleaf pine today is found predominantly on sandy soils, which are less suitable for cultivation. It's not found on the really good soils. Some foresters still think that because they find longleaf chiefly on sandy soils it is a specialist of dry soils. It's a misunderstanding that stretches over at least 150 years. "[The longleaf pine] prefers dry soil and is rarely seen, and never in perfection, on wet or even slightly moist ground," wrote the great Virginia agriculturist Edward Ruffin in 1861. Yet longleaf pine's Latin name, *Pinus palustris*, means "of the swamp," suggesting that longleaf can grow in wet soils as easily as in dry.[12]

"That's the thing that's lost completely, not just to the layman but to ecologists in general," explains wildlife biologist Jay Carter. "In North Carolina, the places you see longleaf now are all sandy soils, where the conditions for the establishment of other things were most extreme. The places you don't see it at all are places with fertile soils, moist soils where other species had a competitive edge once fire was eliminated." He added: "We talk about the 'longleaf pine forest' as if it was one forest type, but you're really talking about a lot of different community types, even in North Carolina. If you're talking about the Southeast, you're talking about a lot more. The vast majority of these are history."

In his Chapel Hill office, Robert Peet confirmed Carter's observation with a graph plotting North Carolina's various longleaf pine communities. The vertical axis showed the percentage of silt in the soil, rising from zero ("sand piles," he said) to high levels of silt and clay. The horizontal axis showed the amount of wetness in the soil, ranging from sandy soils to hydric or saturated soils. If you moved horizontally, you were moving across a gradient from dry to wet, and as you moved up, you encountered soils with more silt.

"On the dry and sandy end, the zero point of both axes, we've got the extremes, deep sand ridges," Peet explained. "If you go toward the wetter sites, but stay in the sandy part of the graph, you first pass through flatwoods and you'll eventually get into the pocosins. But if you move up a little, toward more silt in the soil, now you're running into fire-maintained longleaf pine woodlands, the Piney Woods, as they are called along the Gulf. These aren't the species-rich savannas. They've got a higher concentration of shrubs. Now if you stay on the wet end, but move up toward more silt in the soil here you have an increase in diversity. We've lost shrubs and gained legumes and a lot of other species. It's up there at the wet end that we get the most species-rich savanna systems, the Green Swamp type. This is our concept of the wet savannas."

He moved his finger over toward the top middle of his graph, where the moisture level was lower but the silt levels were high. There was a box with dotted lines. "This is where we've got fine-textured soils but drier. The problem is the communities are gone, extirpated," Peet said. "We don't know what species were in them. They're all missing. You can find fragments on road verges. Follow an old road and as it gets near railroad tracks you may see an area with lots of wildflowers. The mowing has kept these road verges open for the last 100 years. It's only here that you get any hint of what these communities might have been like. Everything else is gone."

FOUR

Webs of Life

*An ordinary citizen today assumes that science knows
what makes the community clock tick; the scientist is equally
sure that he does not. He knows that the biotic mechanism is so
complex that its workings may never be fully understood.*
—Aldo Leopold, *A Sand County Almanac* (1949)

Among the many lessons I was learning about longleaf pine was that the
ecosystem consisted of a host of organisms and natural processes, some of
which played critical roles in the ecosystem. Without fire, for example, the
ecosystem would collapse. But without the grasses, there would be fewer
fires and they would be less intense. Similar relationships exist between
the plants and animals. Indeed, an ecosystem is far more than the sum of
its plants and animals. It consists of the myriad and complex relationships
among them, the web of interdependencies that knit the system together.
Many of these partnerships are vital to the survival of one or more of the
partners, and in some cases they may even be essential to the ecosystem
itself.

How many of these relationships are present in the longleaf pine ecosystem? No one really knows, for these kinds of studies have only recently
begun. Yet piece by piece, scientists are identifying some of the strands of
longleaf's remarkable web of life, and the pictures that are emerging so far
are extraordinary. Among them is the story of the gopher tortoise and its
hospitable burrow, the red-cockaded woodpecker and the tree fungus that
enables it to reproduce, the southeastern fox squirrel and its dispersal of
growth-enhancing fungi throughout the forest, and the amazing and delicate connections between moths and butterflies and their host plants. All
of them tell the same tale about a web of life that evolved over immense
epochs of time.

On an unseasonably hot May morning in 1995 I followed an exasperated wildlife biologist as she moved nimbly through thickets of tick-infested gall-berry bushes on a longleaf pine sandhill site near Gainesville, Florida. Sparse wiregrass languished against a thick duff of needles and oak leaves, enervated by the intense heat that afflicted the region, and crowds of head-high turkey oaks thronged among a few shrubby sand live oaks, a diminutive cousin of the live oak trees that grew along the coast not far away. Rising above the tangle was a thirty-year-old stand of longleaf.

For the moment, the two of us were walking single file, sharing only a few words, pushing past the thick vegetation with our forearms. At 8:00 A.M., we were already sweating profusely. Joan Berish slapped through the overgrown brush, darting glances to the right and left. Finally she stopped and looked around. "Where are all my gophers?" she asked nobody in particular.

Berish wore the uniform of the Florida Game and Fresh Water Fish Commission, the agency she has worked for since 1980. She was looking for light-colored sandy mounds, the telltale sign of the gopher tortoise (*Gopherus polyphemus*), a sandhills inhabitant that she has been researching for two decades. We were in one of two study areas where Berish had worked from 1982 to 1986. During that time she had found nearly 600 tortoise burrows on this site and had captured, marked, and released 178 tortoises. This spring had provided her a chance to see what had happened to the population in the meantime.

The air was still and the conditions were right for an afternoon thunderstorm. Berish had already undergone a rough two weeks in the field, and the threat of another storm was a little unsettling. She talked nervously, intensely. The area was well known for its lightning storms. A few years back, intent on her tortoises, she had ignored the sounds of warning thunder and had been badly frightened when the storm broke fiercely overhead. With lightning striking all around, she clutched three squirming tortoises to her chest—"they weren't stupid; they wanted to get back into their burrows"—and hugged the lowest depression she could find. Last week, her first week trapping tortoises, she had been caught in another storm.

But mostly she was concerned about the missing tortoises. "Only 136 burrows and fifty-one captures in the last eight days," Berish said. "The disturbing thing is that only eleven are recaptures from the 1982–86 trapping season."

I asked her what happened to the rest.

"Well, look at this place," she said, stopping and pointing to the over-grown thickets in which we stood. "In 1983 I had persuaded the landowner here to burn it and he did a good job. It was a hot fire and really knocked the turkey oaks back. But when I came back to start my follow-up study a few weeks ago, I found this. All I can figure is that the lack of prescribed burning has made them move out. It's too thick. Much as I don't want to, maybe I need to go and look at the edge of that giant clearcut that we passed coming in. Maybe they moved there."

She explained that tortoises depended on the presence of the plant foods they ate—broad-leaved grasses, legumes, and even wiregrass. "They're not wild about wiregrass, by any means, but they'll eat it if that's all that's there." Fires, especially fires during the growing season, promoted the growth of this herbaceous vegetation; they also cleared the ground, helping to main-tain the open conditions in which the tortoises like to excavate their bur-rows. In the original sandhill communities, fire occurred perhaps every three to ten years, less often than in the wetter, more mesic sites. But more than a decade without a fire had caused the conditions hospitable to tortoises to worsen.

"It could be we don't know as much about this animal as we thought," Berish said. "Perhaps they're not as sedentary as we thought." An image of lumbering tortoises moving through the thick underbrush in search of open space and a good feed flashed through my mind.

"This is a beautiful sandhill," she said. "It just needs a good burn."

My previous encounters with gopher tortoises had been brief and awk-ward, usually amounting to the sight of its rear end disappearing into a bur-row. When I asked Berish by telephone if I could watch her trap tortoises, I wanted to find out how this animal made its living in longleaf pine sandhills, and how it related to other species in the community.

The gopher tortoise genus *Gopherus* evolved about 60 million years ago in the area encompassed by the Great Plains of today. As many as twenty-three tortoise species may have existed in North America since that time. Four species survive today, three of them in the western United States and Mexico, while the gopher tortoise is the only one found east of the Mississippi River. It's absent from Texas, North Carolina, and Virginia; present in only a few southern counties in South Carolina and a few eastern parishes of Louisi-ana; and concentrated in Georgia, Florida, Alabama, and Mississippi—the middle of the longleaf pine range. It's considered a threatened species in Louisiana, Mississippi, and west of the Tombigbee and Mobile Rivers in Ala-bama. The greatest numbers of gopher tortoises are found today in Florida, where, not coincidentally, there is also the greatest amount of dry, sandy habitat. In Florida, it's a species of special concern.[1]

"Let's check this burrow over here," Berish said, pointing to a mound just beneath a turkey oak sapling. We took a few steps and she suddenly exclaimed: "Hot dog! We caught one!"

A midsized gopher tortoise was helplessly trying to lift its heavy front legs up the vertical surface of a white plastic five-gallon bucket. Berish, her husband, and several others had set the traps two weeks before, placing various sized buckets in holes they dug just outside the burrow entrance, with the open ends of the buckets covered with an oversized piece of brown Kraft paper or newspaper. As the tortoise, which tended to be more active during the courting season, crawled out of its burrow, its weight broke through the paper and it dropped into the bucket. A simple pitfall trap.

Berish pushed a stick into the burrow and gave it a few swipes. Sometimes a rattlesnake was coiled up just inside the cool burrow, she explained. "It would be bad to get bitten on the hand or the leg or foot, but it would be very bad to get bitten on the face."

She bent down, close to the burrow entrance, and hauled the creature out of the trap. I looked at it closely, hoping for a face-to-face encounter, but it ducked into its shell.

Inside its burrow, the gopher tortoise is a fortress within a fortress. When caught outside its burrow, it withdraws into its shell and pulls its huge limbs up behind it like a drawbridge. It has the high-domed carapace of all tortoises but is larger and more massive, averaging from nine to eleven inches in length and weighing about eight to ten pounds, although it can grow much bigger—William Bartram said that the largest specimens he saw had shells eighteen inches long and ten or twelve inches in breadth. It is an extremely long-lived species, reaching ages of forty to sixty years and perhaps exceeding a century. Biologists have found it hard to age these tortoises because the annuli—the concentric rings on their scutes that indicate age—become worn after thirty or forty years.[2]

This is not an aquatic turtle, mind you. Tortoises are strictly terrestrial creatures. The gopher tortoise cannot survive immersion in water, so Berish had drilled holes in all of her bucket traps in case of rain. The tortoise's feet are not webbed; in fact, it has no toes at all, although its stumpy feet have flat nails which, together with its great, flattened forelimbs, are apparently useful for digging in loose, sandy soils. In this environment, the tortoise has learned to be a burrower, commonly digging tunnels about fifteen feet long and six feet deep, although the record is held by a particularly energetic tortoise whose burrow went fifteen feet deep and measured forty-seven feet in length. An astonishing number of other animals make use of tortoise burrows, hence its moniker, "the landlord of the sandhills."[3]

"The first thing I do is check to see whether it's marked or not," Berish said. "This one is not, which means I've got to do it now. Another thing I note is that he's a male. See the way the plastron has a slightly concave shape?" She turned the tortoise upside down to show me its bottom plate. "The theory is that this enables them to climb on the female when they breed." Males have a bigger gular plate, a projection on the plastron just below the head that they use in battle to shove each other around.

Berish quickly measured the length of the tortoise's plastron and carapace with a pair of tree calipers, then suspended him in a sack beneath scales to weigh him. She took a rechargeable drill from her knapsack and prepared to mark him, using an ingenious system by which the various marginal scutes of the tortoise have been assigned numbers — hundreds, tens, and single digits. This tortoise's number was 219, so she drilled the 200 scute, the ten scute, and the two and nine scutes.

"So without further ado," she said as the noise of the small drill filled the morning. When Berish finished she put the tortoise by the burrow entrance and gave him a push. The tortoise didn't budge.

"Sometimes it takes them a few minutes to realize that they're okay and that this big predator hasn't eaten them," she commented. She nudged him again and he disappeared into the hole.

No other tortoise uses burrows as extensively as the gopher tortoise. It remains in a burrow most of the day and may stay for extended periods if it senses danger. One tortoise remained in its burrow for nearly a month before emerging and falling into Berish's trap. The temperature in tortoise burrows tends to be extraordinarily constant — seventy to eighty degrees in summer, sixty to seventy degrees in winter. The gopher tortoise has evolved an instinctive underground life that enables it to survive the fires that frequently run over the landscape and that keeps its body temperature well regulated.

Tortoise burrows are more like communes than single-family dwellings. "Watch this," Berish said as she stirred a mound with a stick. "This is definitely not for the squeamish." Nothing happened at first, and then the burrow seemed to squirm as dozens of small, brown insects began to climb out of the soil.

"Ticks," she declared. "Gopher ticks, *Amblyoma tuberculatum*, the largest ticks in North America, I believe. All gophers have ticks, but I've never seen anything like this burrow in all my years of working with tortoises. Look at them!"

And indeed the burrow was boiling with ticks. I moved back instinctively and pointed out a couple of the pests crawling on Berish's boots. "You can tell they're gopher tortoise ticks because they're trying to get in the sutures of my shoes, as if they were the sutures of a tortoise shell. This might be somebody's idea of a nightmare," she said and laughed.

Wildlife biologists consider the gopher tortoise a critical species in these sandy habitats because its burrow provides shelter and habitat to a bewildering number of amphibians, reptiles, mammals, insects, and even birds, some of which squat in the burrow on a long-term basis whereas others duck in only in emergencies. Sixty vertebrate species and 302 invertebrates—mostly insects—have been found living or sheltering in gopher tortoise burrows. One researcher once saw a cottontail rabbit and a six-lined race runner lizard dart down a gopher tortoise burrow as a fire approached. Other species that have been known to use it for refuge are the endangered indigo snake, the gopher frog, and the Florida mouse, the frog and the mouse both species of special concern in Florida (the mouse is endemic to southeastern pinelands). The mouse tends to construct smaller passages off the tortoise's main tunnel for its own needs and depends on the burrow for its very survival. A number of insects like the gopher cricket and the gopher tick are found nowhere else but in the hospitable confines of the tortoise's burrow. Feeding on the tortoise feces at the bottom of the burrow is a little-studied community of beetles and flies that are preyed on by other creatures that are themselves consumed by the gopher frog and Florida mouse. Some of the insects are not just freeloading; they may actually prey on some flesh flies that parasitize the tortoise.[4]

Tortoises are also reputed to provide other services besides the shelter of its burrows. The mound of soil around the burrow entrance is fertile ground for plants that seed in from elsewhere or are dispersed there by the tortoise's feces. As a result, the mound provides a small island of diversity within the desertlike habitat surrounding it. Its soil is wetter, and the area surrounding the burrow is often the only cleared area in the forest, thus providing protection for seedling plants.[5]

"I've got a tortoise in 131!" Berish exclaimed suddenly as she pushed into a thicket ahead of me. She pulled a small tortoise out of the bucket and checked its plastron. "It's a female, although it's got a pretty big gular. She's not marked. Where are my marked tortoises?"

This tortoise was squirming and trying to get to ground. Its face looked like ET's, or was it merely the familiar melancholy of all turtledom? Its skin was soft as felt. She had two ticks on her face, one on either side, and Berish removed them before beginning her workup. The tortoise's forelimbs were scaly and horny, with the splotchy coloration of chocolate chip cookie dough.

"Their legs are amazingly strong," Berish said. "I've seen a full-grown man trying to pull one out of a burrow and just fall down."

She brought out her drill and started to mark the scutes of the tortoise, which had wisely given up struggling and had withdrawn inside its shell.

"I'm sure she's thinking the worst now," I said.

"Yeah, like why can't you just eat me and get it over with? This playing with your food is for the birds!"

On our way back to Gainesville that afternoon, we talked about the uncertain future that the tortoises were facing. No one knows precisely what the gopher tortoise population is, although most think it is declining rapidly. The tortoise's situation in Florida is especially precarious. With a thousand migrants per week seeking a Florida sunset in their declining years, the biggest threat to the tortoise in the state is urbanization and lack of fire. Predation is a serious problem as well. Before we had left from Gainesville that morning, Berish had shown me an X-ray of a pregnant female she had trapped two days before. In the middle of her stomach were six eggs shaped like ping-pong balls. Statistically speaking, hardly any of these eggs would survive. Once they were laid and buried, most would be dug up and eaten by raccoons, gray foxes, and armadillos. Those that managed to hatch faced an army of other predators, including the reptiles that sought refuge within the tiny tortoise's burrow. In some years, close to 100 percent of the eggs and young are killed by predators.[6]

"The problem with the tortoise is that it doesn't reproduce fast enough to be able to sustain a harvest," Berish explained. "It's not like deer, quail or turkey." She said that tortoises take a long time to reach sexual maturity — thirteen to seventeen years for females in Florida, nineteen to twenty-one years in Georgia. Once mature, they lay only one clutch of four to twelve eggs per year.

At one time the tortoise was hunted as a human food. If you include the Native Americans who no doubt enjoyed eating tortoises, gopher tortoises have been hunted for thousands of years along the Gulf Coast. You could pick up a barrel of tortoises on the Pensacola waterfront in the late 1800s. Truckloads of tortoises were commonly shipped from one spot to another for gopher races.

"Gophers have been popular with a lot of ethnic groups—Cubans and Hispanics down in the Tampa area used to eat them," said Berish. "Gopher pulling was probably started even before the blacks by the Minorcans, people who come from the island of Minorca off Spain and have settled in the St. Augustine area. They have a very strong culture of living off the land and are known far and wide as gopher pullers." One Minorcan told Berish that he and two others pulled 135 tortoises in about seven hours.

"The tortoise was so important to the people around here during the Depression," Berish added, "that it was even called a 'Hoover chicken.'"

As part of her ethnozoological researches, Berish accompanied pullers while they demonstrated their pulling techniques. They would sometimes prepare a dish of tortoises right on the spot using pressure cookers and Coleman stoves. The white pullers liked them with rice or dumplings, whereas the blacks prepared them with a lot of spices. She remembered one telling her: "Ma'am, if you are what you eat, I reckon I'm half gopher." In the black communities, the tortoise still has a medicinal value. "Go pull me a tortoise," someone might say. "I'm feeling puny." Berish gritted her teeth while watching a puller take two female tortoises from a Panhandle site, among the last female tortoises in that area. In the Panhandle the pullers thought she was trying to be highfalutin when she called them tortoises. "Some people think they're a turtle, but they ain't," one puller told her in some exasperation. "They're a gopher."

Gopher tortoises have been a sticky subject in development-minded Florida. Tortoise habitat here and elsewhere in the Southeast has suffered tremendous losses to new residential developments, citrus-growing farms, and other habitat-depriving projects, and there has been a well-intentioned tendency by natural resources agencies to try to save tortoises whose habitat is threatened by relocating them to undeveloped habitat. But without the tortoise, the humble food chain in its burrow—gopher frogs eating dung beetles eating tortoise feces—would be eliminated. This has led some to suggest that the burrow fauna should be relocated as well.

Would the average landowner understand an effort to delay a project until dung beetles and robber flies and gopher frogs and indigo snakes were relocated from a gopher tortoise burrow?

"The average landowner doesn't really have a concept of the ecological importance of the gopher tortoise, and that's real sad," Berish replied. "What worries me and perhaps what is even more insidious is the current backlash against gophers because of development pressures, similar to that facing endangered species all over the country. I wish we had more time to go out to the farm bureaus and spend time with them; we're spread so thin. But wildlifers better start making that a priority or we can lose the whole situation. Unless we can educate these people and engender in them an interest in preserving the habitat, all that we're trying to do may be for naught. We have a monumental educational challenge here."

<hr />

The red-cockaded woodpecker (*Picoides borealis*) is another inhabitant of the longleaf pine ecosystem that is tied in spectacular ways to the forest. It is a small, cardinal-sized woodpecker, about seven inches long, with alternating black and white stripes on its back, a black cap and neck, and white cheeks. The only red in its costume appears on either side of the male's cap, small patches that are visible only when he is defending his territory. The bird chatters like a nuthatch (ornithologist Alexander Wilson facetiously described it as *Picus querulus*). The range of this woodpecker extends beyond that of the longleaf pine, yet it seems to prefer longleaf and, in fact, it is peculiarly well adapted to the tree.[7]

In the original forests, these woodpeckers nested in cavities they excavated in the oldest and largest trees. The cavities are the center of the woodpeckers' lives. Excavating and maintaining the cavities take up an enormous part of woodpeckers' lives, and this work holds the extended family structure together. The red-cockaded is the only woodpecker in the world to excavate its cavity in living pine trees. Most woodpeckers choose an easier route, digging cavities in dead hardwoods or pines and taking no more than ten days and as few as three to complete the job. The red-cockaded woodpecker, however, specializes in living pine trees seventy years of age and older. The bird angles the cavity upward into the tree about six inches, gradually moving through the sapwood and into the heartwood. When it reaches the heartwood, it begins to dig downward, hollowing out a chamber about six to ten inches deep and three to five inches wide.[8]

The requirements of a live pine tree create special problems for the bird. For one thing, it may take several years to dig a cavity. In selecting live over dead trees, the bird has two obstacles to overcome. The first — and the most difficult — is excavating through the sapwood. By digging into the sapwood, the woodpecker triggers a heavy flow of resin that floods the site of the wound and can imprison a hasty red-cockaded in the sticky gum like a fly in amber. Working carefully and waiting for the resin ducts to oxidize and harden typically takes years. The woodpecker's second obstacle in building its cavity is the dense heartwood of the living tree, which dulled the saws of early loggers. Yet this is the easier problem to overcome because resin doesn't flow in heartwood, and also because the bird has a living ally in this process—a fungus.[9]

Foresters consider the red-heart fungus (*Phellinus pini*) to be the most damaging decay fungus to attack conifer trunks. It may extend lengthwise through the heartwood as much as thirty feet, softening it and, incidentally, making it easier for the bird to dig through. Research in the Francis Marion National Forest in South Carolina found that 92 percent of completed red-cockaded woodpecker cavities in longleaf pine were excavated in heartwood infected with the fungus, leading U.S. Forest Service researcher Robert Hooper and his colleages to conclude that "woodpeckers actively select trees with decayed heartwood." Does this mean that the woodpeckers know which trees are infected? The answer may be more a matter of statistical probability than certainty: The birds invariably seem to choose the oldest trees, and red-heart disease tends to infect old trees.[10]

This interrelationship between the fungus, the tree, and the bird is one of the most important symbiotic relationships in the forest. The old longleaf pine tree provides a habitat for the fungus and the woodpecker, and the fungus enables the woodpecker to excavate its roosting and nesting cavity more easily, labor-saving assistance for otherwise labor-intensive and time-consuming work.

This interrelationship also has a profound effect on the social system of the red-cockaded woodpecker.

One July morning before dawn, I accompanied a researcher to a red-cockaded woodpecker colony on a quail plantation in Thomasville, Georgia, where she was monitoring the newly fledged young. The setting was an older longleaf pine forest on rolling terrain that was nearly invisible in the dark. Gradually the eastern quadrant of the forest brightened, and a mist that seemed at first to hang in the treetops slowly settled to the forest floor. The dark shapes of pine trees marched off toward Savannah in ranks, and the brightening sun rouged their edges.

Margaret Sneeringer is one of hundreds of people across the Southeast who have been monitoring the populations of red-cockaded woodpeckers for years. Each morning in the late spring and summer, she rose early and visited a colony of woodpeckers before they dispersed on the day's round of foraging for ants and beetle eggs and spiders. There were twenty-five red-cockaded woodpecker colonies at this quail plantation of several thousand acres, and she would visit each colony site every few weeks to count the birds, note fledgling survival, and identify the family relationships. Earlier in the spring, Sneeringer and other members of Todd Engstrom's research team from Tall Timbers Research Station had fitted each new hatchling with numbered and color-coded leg bands that identified it and its colony affiliation.

Sneeringer set up her spotting scope beneath a tall, flat-topped pine. It was easy to distinguish the cavity trees from the noncavity trees. In the weak, slanting light, the trunks with cavities glistened with resin and were colored a milky green. I focused my binoculars on the entrance hole of one cavity about thirty feet high. Congealed droplets of resin coated the outside of the cavity. Red-cockaded woodpeckers spend a great deal of time drilling scores of small holes called "resin wells" above and around their cavity openings, causing a flow of sticky resin to coat the exterior of the tree. The gum is thought to deter some predators, especially the rat snake, an accomplished tree climber that uses the rough footing of the pine bark to scale the tree in search of eggs or young. Snakes generally avoid the gum.

"I haven't visited this colony in three weeks," Sneeringer whispered. "I found two adults and one fledgling last time."

"Just one?" I asked.

"Before the nesting season there were three adults. The breeding birds laid five eggs. Three of them didn't hatch." She shrugged. "Maybe I missed one. Hopefully I'll catch it today and then in another three or four weeks I'll check them again and continue until September. Over time you can get a pretty good idea of what the population is."

We waited, scanning the trees. The best time of the day to spot them is when they emerge from their cavities in the morning. Once they are all out of their roosts, the colony members usually gather in their area for a few minutes and then disperse across their territory, which could be anywhere from 200 to 300 acres broad. The birds forage most of the day. Males typically forage on the tree's branches and upper trunk, while the females favor the lower parts of the trunk. They return to their cavity trees just before sundown, typically pecking their resin wells for a while to keep them flowing before settling into their cavities for the night. With her spotting scope, Sneeringer was hoping to be able to look at the birds' leg bands and identify them.[11]

"Okay, there's one," Sneeringer said suddenly. She swiveled the scope quickly to the right and glued her eye to the glass. I couldn't see the bird at first and then, there it was, flitting from one tree to another. I followed it with my binoculars only to see it disappear around the back side of a limb. When red-headed woodpeckers forage on a tree, they hurl themselves at the bark with a loud sound display. But the red-cockaded is quieter, more retiring, flicking a little piece of bark with its bill. It lands and then disappears, reappearing minutes later when it flits to another branch and preens momentarily. I caught it again and inspected its black and white dress, hoping it was a male and that it would reveal its tiny red cockade. No such luck.

"Come on, come on," coaxed Sneeringer. She had pointed her scope toward a bird in another tree, but the bird was playing hard to find. She picked up her scope and moved quickly across the open woods and searched the treetop again. For the next hour, we crisscrossed the area, spotting birds and losing them and finally figuring that we had identified all the members of the group. Final count: two adults and one fledgling. It was a grim statistic — of the five eggs laid a month ago, only a single youngster could be accounted for.

Red-cockaded woodpeckers live in family units that are called "groups," and this one was a small unit. A group generally consists of the breeding adults, the young of the year, and several helpers from previous clutches, sometimes as many as nine in all. Each bird has its own cavity for roosting and sometimes nesting; the total number of cavity trees used by the family unit has been called a "colony," although this usage is being edged out in favor of the term "cluster." They are cooperative breeding birds, a trait they share with the Florida jay and about 300 other bird species worldwide. It's a relatively unusual strategy among the 9,000 species of birds. Most animals get on with their breeding rather quickly, but fledgling cooperative breeders enlist as family helpers for a year or more — one red-cockaded woodpecker was a helper for five years — before they become breeders themselves. Delaying their own breeding activity, they help defend the territory, incubate the eggs, feed the young, keep the resin wells pecked, and work on new cavities. All of these activities would seem to benefit the family group and its young but do little to fulfill the individual bird's instinct to breed and thereby contribute its bit of genetic variability to the species.[12]

A small number of male and female red-cockaded fledglings manage to breed in their first year, but this is unusual. Over the last twenty-five years research by Jeff Walters of Virginia Polytechnic University and Phil Doerr and Jay Carter at North Carolina State University has uncovered a variety of roles that fledglings can play. They've discovered that woodpecker behavior seems to be linked to the difficulty in making cavities.[13]

Studying one of the largest extant populations of red-cockadeds, located in the North Carolina Sandhills, the researchers found that most of the young die or are eaten by predators. More females die than males. Almost all the surviving female fledglings will leave the cluster territory and disperse elsewhere, sometimes flying miles to find existing clusters missing females in which they can immediately become breeders. But male survivors, representing 43 percent of the young, have other possibilities. They could leave the territory (13 percent) and become "floaters" or "solitary males," looking for a group in which they could eventually become breeders. They could stay in the cluster and inherit their fathers' territory—only a remote possibility (3 percent). Or, and more likely, they could become "helpers" (27 percent).

But why not disperse, find a mate, and bring up young quickly, thus increasing the number of birds in the population carrying their genes? What's the advantage of delaying their genetic instinct to breed by adopting a subservient position in the cluster? Are these birds inferior to the bolder, pioneering males? Walters, Doerr, and Carter conclude something quite the opposite.

These woodpeckers seem disposed to look for already occupied colonies rather than excavating a cavity and colonizing a new site. Creating a new cavity takes time and energy, for not only must they excavate it, they must defend it from other red-cockadeds or from cavity competitors such as flying squirrels, bluebirds, and red-bellied and red-headed woodpeckers. The surviving young females won't wait around for a year while a young male builds his cavity. They quickly find males that already have territories and cavities. Thus even a bold disperser will most likely put in his time in another cluster as a helper or floater until an opening occurs. And the young, bold ones that do breed in their first year may not be as productive as those that wait. Breeding success increases with age. As a long-lived species, the red-cockaded woodpecker can delay reproduction for several years until a suitable cavity becomes available.

Staying at home for a year and becoming a helper is therefore a viable "choice" for the individual and a critical one for the group. A helper takes turns incubating the eggs and, when they hatch, feeding the young, even after they leave the nest. Of course, life is hard for helpers—out of every 100, 20 will die by the following year while 50 will live to carry on their roles as helpers for at least another year. The remaining 30 have somewhat better luck. Thirteen of them will disperse, while another 17 will inherit a territory from their fathers (nearly a quarter of the breeding males will die each year), but almost all of them will breed. Thus there is a one-in-three chance that the stay-at-homes will become breeders in time, pretty good odds in the bird world.

Walters, Doerr, and Carter showed that staying at home and hoping for a sudden cavity vacancy makes statistical sense. The cavities at home are already completed. Many of the breeding occupants are going to die or be eaten by a predator. Bad things often happen to a red-cockaded family unit, and the stay-at-homes are opportunists counting on the woodpecker mortality rate and the existence of ready-to-use cavities.

"Red-cockaded woodpeckers apparently have existed in a situation so long that they no longer have the ability to rapidly colonize unoccupied habitat," Jay Carter told me. "They look for existing sites. They rarely go out and start new clusters. Their whole social system is based on minimum reproduction—there's no need to produce a lot of young because there's nowhere for them to go. That's true today, but it must have been true for such a long period of time that they developed a social system that's geared that way."

Thus, a lot depends on the fungus that softens the heart of a longleaf pine tree. The survival of the red-cockaded woodpecker species itself largely depends on it. And since this woodpecker is the only species that makes its cavities in the living trees, its labors also benefit the white-breasted nuthatches, bluebirds, woodpeckers, and even wood ducks and screech owls that use the cavities for roosting and nesting.

———

Another relationship in the longleaf pine ecosystem has intrigued biologists studying the southeastern fox squirrel. With its black body and face and its white nose, ears, and feet, the southeastern fox squirrel (*Sciurus niger*) is a handsome tree squirrel, although variations in the animal's coloration are common. At about thirty inches long, from nose to tip of tail, it's the largest tree squirrel in the Western Hemisphere, with a range matching that of the longleaf pine. Four different subspecies of *Sciurus niger*, occupying different parts of the Atlantic and Gulf Coastal Plains, make up what is called the southeastern fox squirrel; another four subspecies comprise the western fox squirrel, a species of deciduous and mixed forests that ranges west from Pennsylvania, across the Appalachians, and all the way to the Rocky Mountains.

Curiously, the southeastern fox squirrel is almost twice as big as its western cousin, a fact that startled Peter Weigl of Wake Forest University and his research associates in the early 1980s. Why was it so big?

Weigl's research suggested that its large size fitted the squirrel to life in the open landscapes of longleaf pine. The other squirrel species of the Atlantic Coastal Plain—the gray squirrel and the southern flying squirrel—are adapted to more densely packed forests in which they can travel from tree to tree while spending very little time exposed on the ground. But in an open, fire-swept woodland like the longleaf pine forest, where the larger trees were often widely spaced, these squirrels found it tough going. They would lose more energy running about on the ground to forage than they could make up through the food they found.[14]

"We found that these very open habitats of scattered oaks and mature pines were ideal for the fox squirrel," recalled Weigl. "And gray squirrels were almost never in there. If the forest was allowed to grow up so that the canopy closed, however, the gray squirrels would come in and eventually fox squirrel numbers would decline."

Weigl's research in North Carolina's Sandhills over nine years suggested that large size benefited the southeastern fox squirrel by enabling it to move efficiently over long distances on the ground. While touring one of the Thomasville quail hunting plantations with a friend one day, I saw a fox squirrel loping along the ground with a half ear of corn in its mouth. It hopped down the path and scaled the first tree it came to, climbing to a branch overhead where it sat flicking its tail. "Where do you suppose it got that corn?" my companion wondered. We had passed a wildlife food plot with several rows of corn at least half a mile away, but could the animal have harvested it there and lugged it all this way?

Yes it could, and no doubt it had. Weigl and his associates estimated that southeastern fox squirrels in their study had home ranges of from thirty-five to fifty-eight acres, the variations depending on sex, season, and food supply. This reflected the patchiness of the squirrels' longleaf pine habitat and their need to range far and wide for food. Western fox squirrels, in contrast, range over two to fifteen acres and gray squirrels' home ranges are about eight acres. For such a large range over such an open landscape, a large body size would certainly be an advantage.

But that may not be the only reason why the animal selected for large size. Fox squirrels eat a variety of foods during the year—everything from underground bulbs, insects, mushrooms, acorns, and pine seeds—but they depend mostly on pine seeds and acorns. The months of June and July are a time of fasting and inactivity for the squirrels because most of their food supplies have been exhausted. Weigl and his associates called this the "disappearance period." The acorns from turkey oaks and other scrub oaks are the squirrels' main food throughout the year, providing 1,000 calories per gram more than a white oak acorn, but the acorns won't begin to drop until October. In late July and August, however, the seeds in the green longleaf pinecones have become ripe enough to be "spectacular energy packages," in Weigl's words. Each one of these large cones can provide as much as 47,000 calories for the animal lucky enough—and large enough—to harvest and process them. The southeastern fox squirrel may well have evolved its extra size while exploiting this food source.

"Only the fox squirrel can manipulate these cones easily," said Weigl. "They're too big for other animals to handle. Gray squirrels can eat them, but with great difficulty. The fox squirrel cuts them green and then carries them off to a feeding station where it chews off the bracts that protect the seeds."

What happens during a poor cone year?

"There are always a few trees that produce even in a poor year," Weigl explained. "That's where the squirrels' size comes in handy. They are able to cover huge areas along the ground and find these trees." Still, a bad cone and acorn year might well leave the squirrels in such poor physical condition that their breeding activity would be disrupted the following year. The researchers could predict the litter size in the spring based on the cone and acorn crops the previous fall.

Fox squirrels also eat various species of hypogeous fungi, or fungi that mature below the surface and produce fruiting bodies in the soil itself. Their reproductive spores are dispersed by animals, mostly mammals. Without their animal dispersers, the fungi could not reproduce. The fox squirrel is adept at finding, digging up, and eating the fungi's subterranean fruiting bodies, eventually eliminating the fungi's indigestible spores in its waste. The big squirrel derives nourishment from the fungus, and the squirrel distributes the fungal spores throughout the forest.

What may be the most important thing about this remarkable relationship is the fungi's importance to the life of the longleaf pine tree and to the ecosystem. These fungi are mycorrhizal. Mycorrhizae are symbiotic associations between the mycelia, or the fine white filaments of the fungus, and the feeding roots of a tree. This partnership helps the tree roots search more efficiently through soil crevices for water and nutrients while the fungus is nourished by the carbohydrates it takes from the tree. Many tree species, including most pines, grow more vigorously with these mycorrhizal associations. Weigl and his associates suggested the interesting possibility that the fox squirrel, the fungus, and the longleaf pine tree have coevolved to take advantage of each other's services.

"The pine provides the fox squirrel with pine cones and a place to live and with fungi to eat," Weigl said. "The squirrel disperses the fungus into bare mineral soil. The fungus needs the tree to live. And the tree not only absorbs moisture and nutrients with the aid of these mycorrhizae, but concentrates minerals in very poor soil. It could be that the mycorrhizal fungi not only contribute nutrients to the plant but may affect the available hormones that would stimulate the rapid growth stage of the tree." In some sandy areas, it is quite possible that the mycorrhizae make all the difference between the pine being able to get out of its grass stage quickly or not. With water and nutrients so critical in such a nutrient-poor setting, the mycorrhizae may provide the tree with enough to survive.

Thus the far-ranging fox squirrel, built for long-distance travel in a fire-maintained landscape, may play a critical role in the forest itself, helping the longleaf pine tree in its adaptation to the environment in which it grows, particularly the droughty, nutrient-poor sites.

―――――

Even more spectacular coevolutionary relationships occur between insects and their plant partners. There are far more insects in a longleaf pine forest—or any other habitat on earth, for that matter—than any other group of animals. If you were to put on a musical hour with the various taxonomic groups of animals taking part according to their relative abundance, the mammals' part of the program would take approximately a minute. The fishes' portion would take about three minutes and mollusks' about five. The insects, on the other hand, would play for about forty-five minutes, with the beetles soloing for nearly half that time.[15]

"In a strictly ecological sense, insects all but own the land," write Thomas Eisner and Edward O. Wilson in *The Insects*. "If man somehow had both the power and was so shortsighted as to extirpate these little animals, the world as we know it would cease to exist. Man himself would probably become extinct."[16]

Thus if you want to penetrate to the inner sanctum of an ecosystem, you must enter the world of insects and their plant hosts. Over the geologic eons, plants and insects have conducted a vigorous and unsparing war with each other. Move about a meadow and a host of flying insects will stir above the grasses—grasshoppers, butterflies, midges, leafhoppers, beetles, bees and wasps. Some of them are sipping nectar from whatever is in flower; others are searching for prey among the teeming multitudes; still others are dispersing away from their natal areas. Unseen are still other insects and their caterpillar young, all of them equipped with mouthparts specially adapted to feed on plants.[17]

In response to these voracious little animals, plants have developed a host of chemical defenses, natural insecticides that we are just beginning to understand. But plants also need to attract the right kind of insects—pollinators and seed dispersers. Plants have evolved a variety of flower shapes and even reproductive strategies, making use of a certain pollinator by flowering in one season rather than another. Insects, in turn, have developed new ways of breaching plant defenses. Thus plants and insects have coevolved over millions of years, each group attempting to get the better of the other, each settling for a dynamic balance of power. "Take away the insects and terrestrial ecosystems would become empty shells within a year," write Eisner and Wilson.[18]

Little work has been done on the insects of the longleaf pine ecosystem, and even less on their plant relationships. One scientist who has is Steve Hall, a zoologist with the North Carolina Natural Heritage Program. Hall has spent more than a decade studying the populations of butterflies, moths, and grasshoppers in the longleaf pine savannas and flatwoods communities of eastern North Carolina, examining especially the effects of fire on their populations. The butterflies and moths provided a perfect focus for the study because they are often tied to a specific host plant during their larval stage and to other plants for adult feeding and mating activities.[19]

I joined Hall for a couple of days at Holly Shelter Game Land, a vast pocosin and longleaf pine wilderness just north of Wilmington, North Carolina. For two hot August days we tramped through the open savannas of this 40,000-acre wilderness, trudged through the deep sands of sandhill communities, and waded through the dense and wet pocosins to reach a ridgetop of old-growth longleaf pines.

Over these two days I slowly became aware of the insect life I had neglected through most of my other travels in longleaf country. I discovered solitary yellow sulphurs flitting lazily from plant to plant, palomino swallowtails, pearly crescents, Georgia satyrs, metalmarks, and buckeyes. There were butterflies so tiny I mistook them for moths, and some so fast I couldn't try out my developing powers of identification. I hunkered down to study lynx spiders camouflaged to match the lavender of an aster. I found a crab spider with a venom so powerful that it regularly attacks and eats bumblebees and butterflies.

I watched Hall as he netted grasshoppers and butterflies. His pouncing movements reminded me remarkably of a fox catching rodents. He sprang forward with a vehemence surprising for the small prey he was after, and sometimes he had to pounce repeatedly to run down a particularly frisky hopper. Hall wore a broad-brimmed canvas hat and he carried field gear on his waist—a large fanny pack for his camera and lenses and two pouches containing specimen jars. He used a microcassette recorder to capture field data, preferring it to writing notes, so there was a constant chatter as we walked, most of it between the two of us but often punctuated by a staccato litany of Hall's recorded observations. "Swarthy skipper perched in grass. First skipper so far," he might suddenly say into his recorder.

In the early afternoon we met Bo Sullivan and Matt Smith, who were subcontracting some of the work that Hall was doing at Holly Shelter. They showed me the results of the moth traps they had set out the previous night. The tray contained a hundred or more moths, a small portion of the thousands that would help Hall chart the distribution of many of these species. I gasped at the size of an imperial moth and listened as Hall explained how important moths were to his study.

"Moths can tell you more about the community you're studying than butterflies," he said. "There's something like 160 species of butterflies in North Carolina, but probably 1,000 species of moths. Only a few butterflies are clearly habitat specialists, but for every butterfly habitat specialist you have ten or twenty species of moths. The butterflies can tell you something, but the moths tell you maybe ten times as much."

That night I helped Hall put out his own moth traps, battery-powered ultraviolet fluorescent lights suspended between plexiglass panels. The lights confuse the moths' navigation system, causing them to fly into the panels and fall into a bucket below filled with the fumes of ethyl acetate that kill them. The next morning I watched as he stored his catch, lifting each small body onto layers of cotton batting. Even though they were not brightly colored like butterflies, their coloration and design were often breathtaking. The moths sat perfect in death, their dull-colored bodies set off against the white cotton.

Later that day we found a colony of yellow pitcher plants sitting in a wet depression, red veining running like livid scars down the flute. The tops of some had turned brown and collapsed, although six or seven inches of the bottom portions of the trumpet remained green. "Caterpillar sign," Hall announced, bending suddenly and stripping the decayed top from one. He pointed to the webbing across the pitcher, where a caterpillar of the *Exyra ridingsii* moth had enclosed itself within its feeding chamber. He peeled the trumpet down its length and pointed to an inch or so of dried caterpillar fecal matter at the bottom that he called "frass." In the late summer the caterpillar had disappeared, perhaps already pupating in its hibernaculum.

A pitcher plant, like other carnivorous plants, is one of nature's excep-tions—a plant that preys on animals. It is a very efficient deathtrap, indeed. In the hollow trumpet of a yellow pitcher plant, you might find several inches of dead and dying prey, from the most recent and living species in the upper layers to the decaying remnants of insects long dead at the very bot-tom of the heap. Beetles, bees, wasps, flies, grasshoppers, crickets, moths, butterflies, even the odd tree frog—all have been tricked into the depths of the plant's inner chamber, and all are serving the grim but practical purpose of nourishing the plant.

But as always, even the most dangerous habitat welcomes the animals that can adapt to them. In the case of the pitcher plant, insects have learned to avoid the traps and exploit the pitchers for their own uses. A certain wasp species builds its nests within them, lays its eggs, and stocks the pitchers with prey for its larvae to feed on. For other insect species, the rotting carrion trapped by the pitcher plant is the attraction. Flesh-fly maggots are satisfied with the decaying bodies at the surface. Midge larvae gorge on what's left. Bacteria breed and dine on the bits and pieces, and protozoans eat them. Mosquito larvae hang upside down from the liquid surface like aquatic vac-uum cleaners, fattening up on all of these. Thus a tiny and, to most of us, invisible, food web is formed, a community dependent on the existence of pitcher plants.[20]

At the very top of some pitchers, shyly withdrawn just below the lip, you can sometimes find a brown moth belonging to the genus *Exyra*. The genus only possesses three species, and two of them carry on their entire lives within separate species of pitcher plant: *Exyra ridingsii* is associated with the tall pitchers of the yellow pitcher plant (*Sarracenia flava*) and *Exyra fax* with the squatter structures of the purple pitcher plant (*S. purpurea*). The third moth species, *Exyra semicrocea*, is a little less selective, feeding on several other pitcher plant species, but apparently excluding other plants.

What is so spectacular about these moths are the ways that each of them has bound itself tightly to the very different structure of its particular pitcher plant, ways that biologist Frank Morton Jones discovered more than a cen-tury ago not far from where Hall and I were walking. The moths lay their eggs within their host pitcher. When the eggs hatch, the caterpillars instinc-tively close up the mouths of their pitchers by spinning thick webs across the top. Protected in this way from predators and to some extent from the weather, the insects proceed to feed on the inside walls of the green pitchers, but not enough to pierce through the walls. In winter, all three caterpil-lars hibernate within their pitcher chambers, but in different ways, dictated in part by the very different structures of their host pitcher plants and the problems each presents.[21]

The pitcher plant community is dependent on fire. Without it, shrubs and other woody plants would shade out the pitchers and they would disappear. Without fire, the pitcher plant community would disappear as well.

Yet fire is also a problem for insects, as I found out later in the afternoon. All day long, a late summer haze had prostrated the fields along the road, silhouetting solitary trees and turning distant woods into dark, shapeless masses. But toward evening, the milky sky had cleared and the strong, slanting shadows of a five o'clock sun striped the open savanna that we slowly made our way across. Purple splashes of *Carphephorus* and *Liatris* enlivened the landscape of waving wiregrass, and goldenrod and boneset added their hues to the scene.

Hall caught a grasshopper in his butterfly net and swung it expertly back and forth to trap the insect within its light folds. He took it between his thumb and his index finger. "*Melanoplus*," he said, "probably *Melanoplus decorus*." The species was yellow-green, only about an inch long, and flightless. Hall pointed to its rudimentary wing and the black line that ran along the thorax, a diagnostic feature of the genus. Removing a bottle from a pouch on his belt, he unscrewed the lid and put the hopper in. It crawled around on a paper napkin soaked in ethyl acetate for a few seconds and then grew still, joining the corpses of five or six other hoppers.

Hall has discovered several Lepidoptera and grasshopper species that are tightly bound to these grassy, fire-swept habitats. *Melanoplus decorus* may well be endemic to the Coastal Plain of North Carolina. One moth species feeds exclusively on the Venus fly-trap and has been found nowhere but North Carolina, while another is tied to pyxie moss (*Pyxidanthera barbulata*), which is found only in the New Jersey Pine Barrens and in coastal North Carolina. A number of skippers have been seen in the Coastal Plain but also in the prairie region of the Midwest. These connections are part of a mass of botanical and zoological evidence pointing toward the existence of similar and perhaps continuous grassy habitat linking these regions to the north and west thousands of years ago.

Hall and Sullivan have also found a number of insect species that are "fire followers," ones that actively seek out burned areas to colonize. One of them, a moth named *Datana ranaeceps*, has been found in great numbers in artillery impact areas at Camp Lejeune, a marine base in Jacksonville, North Carolina. Hundreds of caterpillars belonging to this species have been recovered in areas that are just beginning to green out after a fire. They don't appear on sites that have not been burned for a few years.

All of these species seem to have a couple of things in common, despite their different geographic locations. "A lot of them are grassland species," Hall said as we tried to keep from stumbling over tussocks of wiregrass. "The common denominator is the habitat's openness and the importance of fire. What I've been working on for the last four years is clarifying the interactions between fire and insect populations. It's fairly complex."

"Okay," I said. "Let's talk about fire and insects. Let's take that swallowtail there. A fire comes through, but it's got wings so it flies away. True?"

"Yes," Hall replied. "It could fly away and probably would. But what about its eggs, or its pupae or caterpillars? What happens to them?

"The insect way of coping with fire is a whole lot different than the way that vertebrates and plants cope with it. That's one of the things we're just discovering. Plants have underground root structures. When a fire burns off the top part, they grow back, like the pitcher plants. Or they have very resistant bark, like longleaf pine. Or they have seeds that can remain dormant for a long time until fire clears out an area, and then they sprout like mad. They may get burned, but they survive on site. A woodpecker can fly away. A fox squirrel can climb a tree. They put a lot of emphasis on individual survival. But a flightless grasshopper?" He tapped the killing bottle on his hip where *Melanoplus decorus* was interred. "They don't have a whole lot of options."

Hall told me that what a *Melanoplus* did to cope with fire was confine itself to a habitat that didn't burn very often—the wet ecotone between the pocosin and the savanna or the wet swales out in the middle of the savanna. "It can escape fire by hanging around the wet spots." Some insects apparently escape fire by timing their life cycles to avoid the fire season entirely. Others burrow in the ground, bore into the roots of plants, or feed high in the canopy and pupate underground. The larvae of one rare moth (*Agrotis buchholzi*), for example, hibernate below the soil surface of a few North Carolina longleaf stands before their pupal stage. As caterpillars, they burrow by day and feed at the surface by night, apparently in response to the frequent fires that maintain the moth's single host plant, the pyxie moss.

But most insect species have no such adaptations to fire and no obvious means of escape. Entire populations, not just a few individuals, will be incinerated as a fire—even a surface burn—moves through their habitat. These species deal with fire by recolonizing the burned site from adjacent unburned sites.

"So many insects have developed what is called the metapopulation strategy to deal with fire and other disturbances," Hall said. "Individuals in these species don't lay all their eggs where they grew up. They make a point of spreading out over the landscape, and laying them here and there. The result is you have a lot of little populations scattered over the landscape and a lot of local dispersal between them. So if any one population suffers a decline the site can be easily recolonized." The Arogos skipper (*Atrytone arogos*), for example, is found almost nowhere else but in the fresh growth of new grass that springs up after a fire.

So while a fire stacks the deck against the survival of these species, it doesn't automatically rule it out. A butterfly or moth has a few things going for it in fire-prone landscapes: the adults can fly, they reproduce prolifically, and frequent fire creates a mosaic of unburned patches where populations survive and burned patches where new colonies can be formed.

This survival mechanism apparently worked for thousands, perhaps hundreds of thousands of years, when landscapes were unfragmented and when fires were frequent and patchy. But today, longleaf is confined to smaller areas that fire can easily traverse, Hall noted. "When you had a whole landscape covered in savannas, like you read about in Bartram, you'd always have enough unburnt patches surviving a fire that recolonization would never be a real problem. But now you're restricted to just a few small islands and if you burn them too frequently and there's nothing else outside to serve as a recolonization base, we think you start getting permanent loss of species."

The shadows were growing longer now, and the skippers Hall was looking for were no longer flying. I found a little colony of purple pitcher plants and I knelt to examine their oversized lips covered with backslanted hairs, a slippery chute leading to the lethal pool below. I passed a yellow-fringed orchid with a little grass frog, no bigger than a dime, hanging on the stem. I marveled at its tiny, almost invisible fingers grasping the stalk. *Thwack!* Hall was ahead, swinging his net. I hurried to catch up with him because something was nagging at me, one of those ironies that I was getting used to entertaining.

"Are you saying that even if we reintroduce fire to benefit the ecosystem, we may well be threatening many insect species with extinction?" I asked.

Hall examined another grasshopper, turning it curiously in his fingers. "I guess what I'm saying is a lot of burning strategies that we are using today have been developed with plants, red-cockaded woodpeckers and fox squirrels in mind, and the impacts on the insects are rarely being considered. And there are many reasons why they should be considered."

"Insect pollination, for example."

"Right. And also their place in the food web—birds feed on them, bobwhite quail, turkey poults—they're right there at the base of the food chain. They are also major players in the decomposition food web. It's not just earthworms, you know. A lot of insects are involved in the initial stages of breaking down dead leaf litter. And dispersal—some seeds are dependent on ants to be moved away from the parent plant and actually planted underground. So there are lots of important ecological roles that insects play."

Apparently, if you could ask a butterfly or moth what it feared most in a longleaf savanna, it might say two seemingly contradictory things: "Fire" and "No fire." Either way, if the habitat burned or if the habitat didn't burn, whole populations of moths and butterflies and grasshoppers would perish.

Inwardly I groaned as I realized the enormous implications of what Hall was saying. Managers of a forest like the longleaf pine had more to keep in mind than "just" red-cockaded woodpeckers and gopher tortoises, Venus fly-traps and wiregrass. Their management plans had to encompass even the smallest creatures in the ecosystem, the insects. I was struck immediately by what seemed like the impracticality of it.

I remembered asking Joan Berish that given our present reluctance to make a place for red-cockaded woodpeckers and fox squirrels and gopher tortoises, how in the world were biologists going to convince people that dung beetles, robber flies, and other residents of the gopher tortoise burrows might have a right to be relocated along with tortoises themselves?

"It's hard to believe that people who object to managing for endangered plants or red-cockaded woodpeckers will willingly manage for insects," I said to Hall.

"What we're suggesting is not such a radical thing," he replied. "Mostly what we're recommending is a single rule of thumb: If a landowner is going to burn a single community type like a savanna, don't burn it all at once. Instead, divide it into several burn units. You may eliminate all the insects from the area where you burn but the unburned areas can serve as sources of recolonization. By following this simple policy, we think it will do a lot of good.

"The loss of diversity is something that ought to concern people," Hall added as we walked slowly back to his truck, our eyes blinded by the strong rays of the dying sun. "If we narrow down the list of acceptable species only to those that can tolerate humans we'll end up with rats, cockroaches and house sparrows. It only makes the world a poorer place for humans to live in. I mean, if you're trying to preserve entire ecosystems, doesn't it make sense to consider the impacts of your management on such key elements as the insects?"

The complex relationships in the longleaf pine ecosystem are like underground utility wires, invisible but essential to the very functioning of the system. You don't notice the mycorrhizae growing beneath the surface of a longleaf forest, but they are there providing a service to the trees. You don't see the red-heart fungus within the heartwood of the longleaf pine tree, but the red-cockaded woodpecker depends on it. You hardly notice the insects, but many of them are responsible for pollinating the plants. There is a busy to and fro of animals carrying on their roles as seed dispersers. What happens when you remove a foundation stone from this house that Nature built—the fungus from the longleaf pine tree, the fox squirrel from the forest, the pitcher plant from its tiny but populous community?

When I summed up these adaptations and relationships, I realized what an uncanny resemblance they had to other species within the forest—longleaf pine, wiregrass, and the gopher tortoise, among others. All of these species were long-lived, deferred reproduction over a relatively lengthy time, and produced few young—the classic profile of what some biologists have called "K-strategists." These are species—both plant and wildlife—that inhabit a stable environment in which radical change is rare. The opposite are the "R-strategists," pioneers and opportunists that flood into a rapidly changing landscape and eventually die out or move out in the face of further change.

What did this tell me about the longleaf pine ecosystem? It told me that it was incredibly ancient and stable. That this stability must have been maintained by frequent fires. That over an immense amount of time, large numbers of plants and animals had been able to adapt to fire and to each other in countless ways. That there must be innumerable other relationships between plants and animals that knit the ecosystem together. That what we knew was only a start.

PART TWO

Exploitation

FIVE

Piney Woods People

Having dined at the ferry, I crossed the [Savannah] river into Georgia. . . .
I observed in a high pine forest on the border of a savanna, a great number of cattle
herded together, and on my nearer approach discovered it to be a cow pen; on my coming
up I was kindly saluted by my host and his wife, who I found were superintending
a number of slaves, women, boys and girls, that were milking the cows.
—William Bartram, *Travels* (1791)

Though the intricate adaptations of so many of longleaf's plants and animals to fire and the robust relationships among the forest's web of life seem to indicate an ancient ecosystem, the current forest composition and extent may well be relatively young. Like all ecosystems, longleaf has changed over time, for natural reasons and because of the people who lived in and around the forests.

Studies of pollen in bog and pond sites suggest that climate change has altered forest types and forest composition in the Southeast over the last 30,000 years. During full glacial times, about 18,000 to 20,000 years ago, the vegetation was composed of pine and spruce forests more like the forests of Maine today. During the following millennia, oaks dominated the landscape at certain times and pines at other times. In the warming climate, the pines were eventually victorious and pine forests moved north over the sandy uplands of the Southeastern Coastal Plain, reaching their current northern limits in southeastern Virginia about 5,000 years ago. Many scientists estimate that the longleaf pine forest as we know it today is really only about 5,000 years old.[1]

By then, humans had been living in the region for at least seven thousand years. At one time, historians thought that Native American populations were so low that they could not have had much effect on natural systems. B. W. Wells, for example, wrote that "the Indians were not present in sufficient numbers, nor did they have adequate tools to interfere seriously with the processes going on in the leafy kingdom." That assumption has been shattered in recent decades, with some researchers estimating a much larger Indian presence in the Southeast than was once thought, with possibly profound effects on the natural ecosystems.[2]

At the time of first European contact, 90 to 112.5 million Indians may have lived in the Americas, perhaps 10 or 12 million in North America, and as many as 1.5 to 2 million in the forests of the Southeast. The indigenous people of the Southeast had developed many different language families and cultures over the course of twelve thousand years. Eighteenth-century Europeans generally distinguished among the larger groups such as Tuscaroras, Cherokees, Creeks, Chickasaws, Choctaws, and Seminoles, each of which was dominant in different areas and at different times. Not all of these designations were recognized among the Indians themselves. The Creeks, for example, became a tribal name of convenience for the English in the late seventeenth century, designating all those Indians on the Chattahoochee and the Coosa-Tallapoosa Rivers in Georgia.[3]

Most of these indigenous groups belonged to the Mississippian culture, which arose in about A.D. 800 and was centered on corn agriculture. These Native Americans lived in semipermanent villages and towns along river bottoms where they grew corn, beans, squash, and other crops to supplement their hunting. Hernando de Soto's soldiers constantly encountered Indian towns, cultivated fields, and orchards along fertile river valleys. When he fought his way to Apalachee, near modern-day Tallahassee, Florida, de Soto found fertile fields planted with corn, beans, and squash several miles wide, as well as 250 "large and substantial houses" and a few hundred other dwellings in the immediate area. In Alabama, he passed ten or twelve Indian towns in a single day; in some parts, the cultivated fields stretched along rivers from one town to another. The rise of agriculture among the Indian tribes of the Southeast resulted in increased Indian populations in many areas and new methods of social and political organization based on chiefdoms. De Soto and his men depended on this organization, for they sought out Indian towns to refill their depleted food stores.[4]

Indians didn't live in the pine barrens or attempt to cultivate their sandy soils, preferring the richer soils along rivers or in the mixed pine and hardwood forests, but they used the surrounding longleaf pine woodlands intensively. They collected fallen branches from the forests which they burned for heat and cooking. The resin-impregnated wood burned intensely, and they scraped the soot off the ceilings of their lodges to mix with bear oil for war paint. They used "lightwood" torches that burned brightly at a mere touch of fire in their hunting and fishing. Indians paved their village streets with pine bark, and they used pine posts, "which will last for several ages," as frames for their summer houses, according to English Indian trader James Adair.[5]

Indians burned the forests, too, although this practice was not confined to the Southeast or to the Coastal Plain forests. Travelers' accounts up and down the Atlantic Coast often contain references to the natives' custom of burning the woodlands each year. Their fires encouraged browse for deer and other animals, and they made hunting easier by opening the woods. Fire opened new areas for crops and maintained old fields. Indians used fire to girdle trees; gather honey; control ticks, chiggers, and other vermin; and in warfare and signaling. They were not too concerned about putting fires out, and so their fires, like natural fires, could burn extensively. This was potentially a lot of annual burning on top of the lightning-ignited fires that burned every few years. Might these annual fires have created the open woodlands that Europeans observed in the Southeast?

Indeed, despite their differences about whether the forest was monotonous or diverse, Europeans generally agreed that longleaf pine forests were amazingly open. The "country is so good, that one may ride full gallop 20 or 30 miles on end," said an early traveler in Georgia. Sir Charles Lyell, an English geologist who explored the South in 1841 and 1842, recorded a visit in the North Carolina Piney Woods during which he had "many long rides together through those woods, there being no underwood to prevent a horse from galloping freely in every direction." The pine trees were so far from each other, wrote another visitor, that they gave the appearance of "open groves rather than of forest."[6]

William Bartram invariably characterized the pine forests as "open," "airy," "thin," or "scatteringly planted;" interspersed among them were treeless or almost treeless places that he called "savannas" or "meadows." These were emotionally contrasted with the dense, closed-canopy hardwood forests—"gloomy forests," "awful forests," "frightful thickets"—that he encountered from time to time. Bartram described his encounters with open savannas and prairies on almost every page. The word "savanna" was a Spanish import, borrowed from Indians in the West Indies; it meant a flat, grassy, and treeless or almost treeless area. Bartram employed an endless and seemingly effortless vocabulary of adjectives to portray them. They were "large, rich savannas," "extensive savannas," "illumined savanna," "unlimited savannas," "grassy savannas," "extensive, level, hard, wet savanna," "embroidered savannas," "extensive green meadows or savannas," "endless savannas," and "delightful green savannas." They were such a common sight in North Carolina, for example, that Edmund Ruffin defined them as a distinctive landscape feature: "interspersed among, and surrounded by forest, tracts of open grass land . . . as clear of trees as if made and kept so for cultivation." He noted, too, that savannas could either be treeless or sparsely covered with pines. "Indeed, in many parts of the pine lands, where the trees stand but thinly, the savanna grasses cover the ground, and where burnt off, present the appearance of a true savanna, with the addition of a sprinkling of trees." Another traveler compared them to "isolated prairies on a small scale, covered with a tall, rank grass."[7]

"Open and parklike" forests became somewhat of a cliché among many English settlers, whether in Massachusetts or Georgia. We sometimes think that early settlers met with a dense, closed-canopy forest that stretched for a thousand shadowy miles all the way to the western prairies, but settlers encountered remarkably open areas and grassy prairies long before they lost sight of Plymouth Bay, Charles Town harbor, or Savannah. Fire historian Stephen Pyne suggests that these open pine woods and treeless and nearly treeless savannas may well have been the result of countless annual Indian fires, and that these fires were a potent factor in shaping and even expanding the environments that Native Americans preferred. "The dominant vegetation type in America [at the time of European contact] may well have been grassland or open forest savannah," he writes.[8]

Other openings noted in southern pinelands were old town and field sites abandoned by Indians. Their corn agriculture rapidly depleted the soil of nutrients, and they tended to exhaust the supply of wood from the nearby forests. As a result, Indians tended to move every so often, leaving behind the cleared spaces of their towns and fields that the Creeks called "tallahassees." The extensive Ocmulgee Indian town and mounds outside of Macon, Georgia, were home to about one thousand Indians in about A.D. 900. When Bartram visited the site in the 1770s and paid homage to "the power and grandeur of the ancients," the town had already been abandoned by the people who built it for more than six hundred years.[9]

Still other openings in the forests were melancholy reminders of just how quickly the Indians melted from the landscape. When the English were settling Savannah, southeastern Indian populations had been in the throes of a spectacular decline for more than a century. By 1700 Indian populations in the Southeast may have declined by as much as 90 to 95 percent from their population levels at settlement, although this figure is somewhat controversial. In 1775 Indian trader James Adair observed that the Indian nations were declining rapidly "on account of their continual merciless wars, the immoderate use of spiritous liquors, and the infectious ravaging nature of the small pox."[10]

Europeans were responsible for the spread of liquor and disease among the Indians, and the "merciless wars" that Adair mentioned were a direct result of the power struggles among the French, Spanish, and English who employed Indians as mercenaries and slavers. The English in the Virginia and South Carolina colonies were particularly adept at setting one tribal group against another, offering guns, cloth, and other goods in exchange for young men to be sold as slaves to plantation owners. The effect on Indians was devastating. Depopulated by disease and constant war, Indians abandoned ancestral lands and relocated, often hundreds of miles away. Fragments of different tribes coalesced into new groups with new names and sought firearms for protection and also to participate in the market for slaves and deerskins themselves. The Yamassees in coastal South Carolina, for example, became notorious in 1715 for their bloody uprising against the English, but they probably didn't even exist in 1600, much less decades earlier when de Soto had first passed through.[11]

Ironically, by the time James Oglethorpe was building his palisade around Savannah in 1734, relatively few Indians remained in the forests, mostly detribalized elements who were even more afraid of the slave-hunting Yamassees than Oglethorpe himself. Northern Florida had undergone desperate changes in the two centuries since de Soto's expedition. Attacks by English marauders allied with Creek slavers in the early eighteenth century had devastated the Apalachee region, described by Cabeza de Vaca and de Soto as fertile and well populated. Many of the Apalachee people had been killed or enslaved on the rice plantations of the Carolinas, with survivors of the raids trekking west to escape. Later in the century some of them drifted back, joined by refugee Yamassees from South Carolina and even descendants of the original Creek raiders. Bartram, traveling through northern Florida from St. Johns River to Tallahassee in 1774, described the Indians as happily herding their cattle and tending their crops. But they were not the original Timucuan society, once 200,000 strong throughout north-central Florida and southern Georgia, nor the Apalachees, perhaps 50,000 strong in the region around Tallahassee at the time of the Spanish invasion, but a patchwork group of refugees and descendants of slavers who had resettled the region and whose culture, as the presence of cattle might indicate, was vastly changed.[12]

By the late eighteenth century Indian remains were nearly as remarked upon as the Indians themselves. William Bartram and travelers in Georgia and Florida constantly noted the evidence of past Indian occupation — "ancient Indian towns," "magnificent monuments," "conical mounds," "old Indian settlements, now deserted and overgrown with forests," "ancient Indian fields, where there are evident traces of the habitations of the ancients." Even the Spanish had disappeared, leaving their own "vestiges" and cultural "monuments" to mark the land and even the Indians themselves. The old 200-mile *camino real* from St. Augustine to Apalachee Bay, which connected the Spanish missions like pearls on a string, was grown up in many places with trees and shrubs. Many of the Seminoles in Florida were Christians, and most spoke and understood Spanish. They herded cattle that they had stolen from Spanish haciendas and rode horses whose veins ran with the blood of Arab mounts that had swept through Europe centuries before.[13]

———

Besides creating an open forest, could annual Indian fires have had even more profound effects on the forest—altering the evolution of forest organisms, for example, or the flowering time of vegetation, or the composition of the forest itself?

It's a difficult question to answer, as I discovered one day while touring the Wade Tract Preserve in Georgia with Ron Myers, the director of fire management for The Nature Conservancy, and Sharon Hermann, a research plant ecologist who was then working at Tall Timbers Research Station. I wondered, somewhat idly, whether there was a "natural" season of fire. Didn't the presence of humans in longleaf pine forests with their own fire patterns complicate things? Naturally occurring fires during the growing season may well have shaped the evolution of plant and animal behavior in longleaf ecosystems, but Indians and Europeans burned the woods at other times. At Florida's Choctawhatchee National Forest early in the century, turpentiners tended to burn the woods in the winter, cattlemen burned them in early spring, and hunters burned them during the dry months of autumn. And, of course, the Indians burned as well, over a much longer period of time.

So is there a natural time for the forest to burn? I asked.

"It depends on what you mean by 'natural,'" Myers answered warily, darting a look at Hermann. "There are many people who think that a suite of species evolved under a lightning fire regime before humans were here. But I'm not sure that means that you should manage longleaf pine forests today under that regime. That regime occurred, who knows, maybe 20,000 years ago when the climate was very different. The longleaf pine–wiregrass community or the longleaf pine–bluestem community may have been much more restricted in extent. Maybe it was restricted to ridgetops and xeric areas subject to frequent lightning strikes.

"But we know that within the past 10,000 to 12,000 years, there were Indians here, doing their own burning, probably in the fall and winter. So superimposed over 'natural' summer lightning fires you had Indian-set fires. The extent of the longleaf pine forest—from Virginia to Texas—may well be an artifact of Indian burning."

I pondered this in silence for a moment. When the Indians invaded the Southeast, glaciers still encased much of the North and boreal forests of hemlock, spruce, and jack pine covered extensive areas of the Southeast. Longleaf pine may well have been forced well to the South, down the Florida peninsula, perhaps, onto the exposed continental shelf or even into Mexico. As the climate warmed and lightning-ignited fires increased, longleaf pine and its associated grasslands began to expand, a process that generations of Indians must have witnessed and possibly even abetted by burning the grassy pinelands.

Hermann wasn't convinced. "Ron and I are going to differ over this a little bit," she said finally. "I'd say more that it was only possible that the ecosystem was human-altered. I guess I just don't have a really good feel for it."

"She doesn't have any faith in the Indians, that's what it is," Myers said with a grin.

"No, what I don't have is a particularly solid idea how widespread at any one time the Indians were. Depending on the density of the population, that will influence how frequent the fires that they started were. But I'm not sure it's worth arguing about because it's not resolvable—I just don't know how you'd research it. For me it's just pretty hard to say that Indians influenced the assemblage of plants and possibly the evolution of particular plants."

Descriptions of the Piney Woods began to appear in travel accounts of the late eighteenth century and continued after the defeat of the South in the Civil War. Many of these accounts disparaged and ridiculed the people of the Piney Woods. In North Carolina, Frederick Law Olmsted considered them "entirely uneducated, poverty-stricken vagabonds . . . people without habitual, definite occupation or reliable means of livelihood." Edmund Ruffin, traveling through coastal South Carolina, judged the inhabitants of the vast pine woods to be "a most wretched & worthless population." Said another: "They wander in the woods and hunt and fish because they lack the capacity to live in any other way. . . . They despise labor because it is wearisome, and their repugnance to it is only conquered by a fine prospect of shooting deer, or cutting a bee-tree and filling his buckets with the snowy combs of wild honey." The Piney Woods were the home of "crackers" and "poor white trash."[14]

Not everyone despised them. J. F. H. Claiborne admired them for "living in a state of equality, where none are rich and none in want; where the soil is too thin to accumulate wealth, and yet sufficiently productive to reward industry, manufacturing all that they wear, producing all they consume; and preserving, with primitive simplicity of manners, the domestic virtues of their sires." Yet in a region where settlements were often twenty miles apart, even Claiborne found it lonely. Other writers blamed the people's coarseness on their isolation. "A solitude of unsettled forest rapidly barbarises," wrote one. "Even the educated and refined, placed in the same circumstances, become gluttonous and careless of dress, and give up to coarse and slothful manners."[15]

Some contemporary historians believed that these depictions misrepresented the Piney Woods people. Olmsted in particular comes in for criticism for mistaking the rough-hewn pioneer appearances for reality. Thad Sitton, in *Backwoodsmen: Stockmen and Hunters along a Big Thicket River Valley*, says that the "poor white trash" Olmsted described were not lazy farmers but *stockmen* whose wealth—vast herds of cattle and hogs that grazed on the open range—was invisible to Olmsted. The sheer numbers of cattle and hogs in the Atlantic and Gulf Piney Woods explain why the settlements were so isolated. The people were herdsmen who preferred areas of poor soil that offered excellent forage for cattle and hogs. "They prefer those extensive pine barrens, in which there is such inexhaustible range for cattle, and which will not for a long time admit a dense population," wrote historian Timothy Flint in 1832.[16]

Indeed, European settlers discovered almost at once that the open pine forests of the Southeast had immediate uses as pastureland for their domestic animals. In his brilliant study of the origins of cattle ranching in North America, Terry Jordan found that the herdsmen of the longleaf pine country in the Carolinas and Georgia developed many of the cattle ranching traditions we normally associate with the western plains. Under the nearly continuous cover of longleaf forests, cattle ranchers carried these traditions from the open savannas and pine woods of the southern Coastal Plain all the way to Texas.[17]

The early settlers in the Carolinas and Georgia grew their crops in the fertile soils of the bottoms, following the Indian custom. At first, herders penned their cattle and hogs at night, but that became impractical as the numbers of these animals grew larger. Eventually the people let the livestock roam unpenned while fencing their gardens and crops. The open pine woodlands, savannas, prairies, swamps, canebrakes, and hardwood forests were an excellent blend of habitats for the animals' needs. In spring and summer cattle browsed on the greening grasses and shrubs of the pine barrens, whereas in fall and winter they moved into the swamps and canebrakes to feed on the young leaves and tender shoots of cane. In time, the herders, following Indian practices or their ancestral traditions from Ireland or England, burned the woodlands and savannas in winter to create new grazing while the cattle foraged in the bottomlands. Cowpens, or fenced enclosures to make it easier to brand or mark cattle for market, were built in the treeless or almost treeless savannas.

The cattle required such little management that vast herds could be easily amassed. "It is nothing uncommon for one man to own 100 or more head of horned cattle; some count their heads by the thousand, all running loose in the woods and swamps," wrote German traveler Johann David Schoepf about eastern North Carolina in the 1780s. Cattle herds of 1,500 to 6,000 head ranged between the Ogeechee and Savannah Rivers. The cattle were thin and half wild, but they cared for themselves. Mounted cowboys using long whips drove the cattle to Charleston, where there was a ready market for beef.[18]

By the time Schoepf wrote, cattle ranching in the Southeast was already 250 years old. The Spanish raised cattle in Florida as early as 1521, and by 1700 there were thirty-four large ranches in Florida that had as many as twenty thousand head of cattle. By the time Bartram visited them, the Seminoles of Alachua were accomplished horsemen and cowboys, driving their herds along established trails. In a short time, cattle had become so important among the Seminoles that they bestowed honorific "Cowkeeper" upon their chief.

New arrivals in the backcountry of the Carolinas and Georgia pushed the herdsmen to move to similar pasture to the south and west. They migrated from the Atlantic Coast to the Piney Woods of Florida, Alabama, Mississippi, Louisiana, and East Texas, a range that was practically identical to the one they had known in the Carolinas—open pine woodland, sparsely treed savannas, and a mix of swampland and hardwood forests. In 1841 J. F. H. Claiborne described the customs of cattle herders in Greene County, Mississippi, in the midst of extensive forests of longleaf pine:

> Left Augusta for Mr. Bruland's, a very comfortable house of entertainment some sixteen miles distant. Passed on next day through a level open pine woods country to Leaksville, the county seat of Greene. . . . Many of the people here are herdsmen, owning large droves of cattle, surplus increase of which are annually driven to Mobile. These cattle are permitted to run in the range or forest, subsisting in summer on the luxuriant grass with which the teeming earth is clothed, and in winter on green rushes or reeds, a tender species of cane that grow in the brakes or thickets in every swamp, hollow and ravine. The herdsmen have pens or stampedes at different points in the forest, where at suitable times they salt the cows, and once or twice a year they are all collected and marked and branded. This is a stirring period and quite an incident in the peaceful and somewhat monotonous life of the woodsman. Half a dozen of them assemble, mounted on low built, shaggy, but muscular and hardy horses of that region, and armed with raw hide whips of prodigious size, and sometimes with a catching rope or lasso, plaited of horsehair. They scour the woods in gallant style, followed by a dozen fierce looking dogs; they dash through swamps and morass, deep ravines and swim rivers, sometimes driving a herd of a thousand head to the pen, or singling out and separating with surprising dexterity a solitary steer which has become incorporated with another herd. In this way, cheering each other with loud shouts and making the woods ring with the crack of their long whips and the trampling of the flying cattle, they gallop thirty or forty miles a day and rendezvous at night at the stamping ground.

It's probable that most of the herders that Claiborne observed, or their fathers, had moved to Mississippi from the Carolinas and Georgia not long before. "So rapid was the westward thrust that a man who had learned the cattle business in colonial South Carolina could have ended his days raising cattle in Louisiana or even Texas," remarks Jordan in *Trails to Texas*.[19]

Developed in the continuous longleaf pine forests of the South Atlantic Coastal Plain, the sixteenth-century herding practices later merged with the Spanish customs of Florida and Mexico to produce the more well-known cowboy traditions of the western plains. Open-range cattle grazing in the

Southeast was colorful, but it had negative effects on the longleaf pine forest that could be felt well into the twentieth century. Early twentieth-century postcards of cattle-grazed longleaf pinelands show a close-cropped ground cover that was not at all natural. Jordan says that excessive numbers of cattle trampled and compacted the soil, destroyed the diverse, perennial vegetation, and encouraged annual species. Worse, by reducing the presence of native perennial grasses, cattle grazing reduced the fuel for fires that were necessary to perpetuate the forest.[20]

The other domestic animal that made use of the Piney Woods—the hog—was introduced by de Soto in 1539, and generations of later settlers added their own. Like cattle, hogs ranged freely in the plentiful forests and swamps of the South. They ran in large herds and reproduced rapidly—the thirteen that accompanied de Soto had grown to seven hundred by the time he reached the Mississippi River. Sows breed when they are six to nine months old, and they can raise two litters of four to eight pigs each year. In the wild, they became lean and agile, with fighting teeth and a high back that earned them the name "razorback."[21]

Over the following centuries, settlers let more and more hogs forage in the woods, gathering them up when it was time to drive them to market. They used specially trained hog dogs to hunt them or used stacks of cornstalks to entice them into pens, where they were fattened with corn for five or six weeks. Some hogs weighed as much as five hundred pounds when they were driven to market.[22]

The volume of hogs and cattle in the Southeast in general, and in the piney woods in particular, was startling. The 1850 U.S. Census recorded just over 6 million people, 6,306,509 cattle, and 11,872,826 hogs in the southeastern states. In North Carolina, Georgia, Alabama, Florida, and Mississippi there were more than two hogs for every person; in Texas there were more than three.[23]

What William Bartram saw was a land shaped by natural forces, to be sure, but it was also a land that bore the imprint of a human presence already twelve thousand years old. Yet Indians had a milder effect on the longleaf pine forests than the Europeans. European livestock changed the forest, and in the centuries following, tar burners, turpentiners, lumbermen, and foresters would exploit the forest to an even greater extent, all of them unwittingly altering it in ways great and small. Native Americans had lived as an integral part of the longleaf ecosystem for thousands of years, but less than four hundred years after European settlement the forest had for all intents and purposes disappeared.

SIX

Tar Kilns and Tar Heels

Brer Fox went ter wuk en got 'im some tar, en mix it wid
some turkentime, en fix up a contrapshun w'at he call a Tar-Baby.
—Joel Chandler Harris, *Uncle Remus, His Songs and Sayings* (1881)

In 1528 Spaniard Pánfilo de Narváez led an army of three hundred conquistadors up the Gulf Coast of peninsula Florida to Apalachee, near present-day Tallahassee. The expedition ended disastrously. Miserable, ill, hungry, and beset by hostile Indians, they returned to the coast after a few months, ate the last of their horses, and set sail in makeshift boats, hoping to return to Havana. They didn't know how to make boats, wrote Alvar Núñez Cabeza de Vaca, one of the few survivors of the campaign, nor did they have naval stores—neither "iron, nor forge, nor [oakum], nor pitch, nor rigging." They made do. One of their members was a carpenter who used palmetto fibers and husks to make an adequate substitute for oakum, normally a tar-covered hemp that carpenters pounded into the seams of ships to prevent leaks. Another, a Greek named Don Teodoro, made pitch from "certain pines" to waterproof their hulls. Despite these efforts, only three men survived the return trip. After eight years of wandering, the others were dead of disease and starvation, drowned in rivers, riddled by armor-piercing Indian arrows, or cannibalized by their friends.[1]

Their expedition is notable because it was probably the first documented European use of naval stores on the North American continent and certainly the first use of southern pines for that purpose. Naval stores have slipped from view today, but throughout history nations have depended on them, sought them out, fought wars over them, and treated them like nations today treat petrochemicals. Without them, and without access to the forests from which they came, a nation's military and commercial fleets were useless and its ambitions fruitless. Until the nineteenth century, the greatest navies in world history were kept afloat by the humble products of conifer forests.

Although the Spanish were generally indifferent to the forests they passed through on their search for El Dorado or the Seven Cities of Cibola, the English weren't. When the first English explorers viewed the vast forests of pines on the southeastern Atlantic Coast, they sensed immediately what we know today—that the pine forests of the Southeast were the greatest naval stores forest in the world. The English saw trees, as Philip Amadas and Arthur Barlow reported to Sir Walter Raleigh in 1584, "trees which could supply the English Navy with enough tar and pitch to make our Queen the ruler of the seas." The English didn't waste any time. Just a year after the founding of Jamestown, tar burners were on their way to Virginia carrying a technology that was already thousands of years old.[2]

When and where people first exploited pine trees for their resinous properties is unknown, but no doubt it occurred in a maritime country when sailors first dared to set forth on the open seas in a vessel made of wood. In the ancient Mediterranean world, Greeks, Macedonians, Phoenicians, and others tapped a variety of pine trees for resin which they cooked in open cauldrons. This primitive distillation process created an oily vapor that shipbuilders captured on sheep skins and wrung out to make a form of turpentine spirits. They called the black, concentrated gum left behind pitch. Greek shipmakers caulked the seams of their triremes with pitch, and they smeared the hull with pitch to seal it against leaks. In the Homeric poems, the hulls of the Greek ships are described as black with pitch; only the bows had patches of color. Before Noah's great adventure, God instructed him to "make yourself an ark of gopherwood. Put various compartments in it and cover it inside and out with pitch."[3]

The trees needed to make pitch were not equally distributed in every maritime country, so pitch became a critical part of ancient commerce. If you were building a fleet of grain carriers in Italy in the third century B.C., you might get the main building timbers from the region of Mt. Etna and the wood for the tree nails and frames from Sicily. From Spain would come the esparto or grasses that made your ropes, and from the Rhone Valley in Gaul came the pitch for your hull.[4]

From ancient days until the end of the wooden ship era, the smell of tar and pitch was forever mixed with the sea's salt air in the memories of sailors. If Aristophanes associated the fitting out of a ship with pitch in the fifth century B.C., John Muir was still referring to "the Tar-scented community of the ship" twenty-five centuries later. By then, the words "tar" and "pitch" referred to two distinct conifer products: tar was derived by slow-burning pine wood in a kiln, while pitch was made from tar boiled in a kettle or cauldron. By the seventeenth century the maritime associations of pine derivatives had bestowed upon them the term "naval stores." These were the commodities required to build and maintain wooden ships, and they included not only tar and pitch, but also spirits of turpentine and rosin, derived from tapping the gummy resin from live trees. Other naval stores included planking, masts, canvas, ropes, and tree nails. No vessel could be built without them, and scores of maritime occupations depended on them.[5]

In the bustling shipyards of long ago, barrels of tar and pitch were everywhere. When the ship framers had completed their work and had planked the ship's ribs, the caulkers went to work. Using mallets, wedges, and caulking irons, caulkers drove oakum and cotton into the seams between the planks to make the ship watertight. Oakum was tarred hemp fiber that was shipped in great bales and broken down by the caulkers into smaller balls which they hammered into the seams. Then the shipbuilders coated the underside of the hull with pitch, making it watertight and protecting it from the dreaded and destructive *Teredo navalis*, the tiny wood-boring mollusk. To protect against the *Teredo*, shipbuilders often sheathed a ship's hull with a mixture of tar and hair and nailed half-inch boards to cover the planks.[6]

Because ships at sea were exposed to corrosive salt air and saltwater for months and sometimes years at a time, ropemakers treated ropes or cordage with tar as a preservative. A ship possessed miles of ropes, of many different diameters ranging from a quarter inch to eight inches or more, depending on the size of the ship. Some belonged to running rigging, or the movable ropes that operated the sails. Others were part of the standing rigging, the heavier, permanently attached ropes that kept the masts in place. Hawsers moored the ship; massive cables raised and lowered the anchors. The ropemakers tarred all of them, most of them "in the yarn," meaning that the individual yarns received a thin coating of tar before they were fashioned into rope.[7]

Making cordage was a skilled occupation. For large merchant ships or military vessels, ropemakers carried on their trade in the ropewalk, a long building in the shipyard where they combed out hemp fibers and spun them clockwise to make yarn. Workers carried the yarn into the tar house, where it was coiled into a kettle of boiling tar and slowly drawn out by a capstan. The tar needed to be hot enough to penetrate the yarn, but if it was too hot it wouldn't stay in the yarn. Similarly, the yarn needed to sit in the tar long enough for the tar to penetrate, but not so long as to tenderize it. Yarn intended for cables required more tar than for hawsers; hawsers needed more tar than standing or running rigging. The tarred yarn was then twisted counterclockwise to make strands, and the strands were twisted clockwise to make rope. Twisting tarred strands into ropes of uniform diameters required wheels to generate the necessary pressure.

Once the standing rigging was in place on the ship, the riggers tarred them again. Margherita M. Desy, associate curator of the USS Constitution Museum in Boston, says that this not only waterproofed them, but also made them stiff. "This was important because the crew had to climb aloft," she said. "If the rigging was too flexible, they wouldn't be able to climb or they would climb more laboriously. But if you made sure your rigging was taut and then tarred it, that would stiffen the rigging and a sailor could climb it like a ladder. He literally could run up a tarred rigging."[8]

Tar and pitch were essential not only in the construction and rigging of a ship, but also in its maintenance at sea. Before setting out on a voyage of any length, every ship was outfitted with provisions and stores. Detailed inventories of these outfits survive from the wooden-ship era to suggest the needs of sailors aboard wooden ships at sea. In 1781, for example, the Continental Frigate *Alliance*, carrying twenty-eight 12-pound guns and eight 9-pound guns, set sail from Boston for the port of L'Orient, France. From its inventory we know that on the voyage the sailors ate both fresh and salt beef, pork, bread, vegetables, and cheese; drank rum (both New England and West Indian varieties) and coffee; and seasoned their drink and food with butter, sugar, salt, and molasses. Under the care of the carpenter, armorer, gunner, cooper, surgeon, and master, she carried nails, topmasts, sails, ropes, swivel guns, hand grenades, muskets, hammers, sheet lead, handsaws, syringes, writing paper, "Salt Sellars," candlesticks, and other stores. The boatswain was responsible for two and a half barrels of tar and thirteen tar brushes. Among the carpenter's stores were two barrels of pitch, three barrels of tar, and two barrels of raw pine gum.[9]

In 1844 a three-deck ship of the line carried six barrels of tar on a one-year voyage, nine barrels on a two-year voyage, and twelve barrels on a three-year voyage. The smaller brigantine on a one-year voyage required a single barrel of tar, whereas two barrels were allowed for both two-year and three-year voyages. A whaling ship, making a three-year voyage around Cape Horn in 1874, on the other hand, needed three barrels of tar.[10]

At sea, these supplies would be used for routine maintenance as well as for emergencies. In a standard procedure known as "tarring down" the rigging, sailors periodically climbed to the mast tops and smeared tar from top to bottom. Richard Henry Dana's 1840 narrative, *Two Years before the Mast*, recorded this vivid description of the process:

> We put on short duck frocks, and taking a small bucket of tar and bunch of oakum in our hands, went aloft, one at the main-royal masthead, and the other at the fore, and began tarring down. This is an important operation, and is usually done about once in six months in vessels upon a long voyage. It was done in our vessel several times afterward, but by the whole crew at once, and finished off in a day; but at this time, as most of it, as I have said, came upon two of us, and we were new at the business, it took several days. In this operation they always begin at the masthead, and work down, tarring the shrouds, backstays, standing parts of the lifts, the ties, runners, &c., and go out to the yardarms, and come in tarring, as they come, the lifts and foot ropes. Tarring the stays is more difficult, and is done by an operation which the sailors call "riding down." A long piece of rope . . . is taken up to the masthead . . . into which the man gets with his bucket of tar and bunch of oakum; and the other end being fast on deck, with someone to tend it, he is lowered down gradually, and tars the stay carefully as he goes. There he "swings aloft 'twixt heaven and earth," and if the rope slips, breaks, or is let go, or if the bowline slips, he falls overboard or breaks his neck. This, however, is a thing which never enters into a sailor's calculation. He only thinks of leaving no "holidays" (places not tarred)—for, in case he should, he would have to go over the whole again—or of dropping no tar upon deck.[11]

Sailors were called "tars" after the word "tarpaulin," a tarred canvas cloth. English sailors were known as "Jack Tars." "Tars" had a noble ring, but it could also be used contemptuously as when sailors were called "tar-breeches" and "tar-lubbers." It may be no surprise that the punishment of being "tarred and feathered" originated at sea—it was ordered by Richard I of England as a standard naval punishment for theft, although it also had plenty of applications on land. The literary uses of the term continue to our own day. Stephen Maturin, the hero of Patrick O'Brian's popular series of sea novels, derives his last name from *mathurin*, the French word for Jack Tar.[12]

When a ship sprang a leak at sea, not an altogether uncommon event, sailors caulked the seam using oakum and pitch. Sometimes ships had to be intentionally grounded, especially if it was the "devil seam" that had sprung the leak. This was a particularly troublesome seam because it lay at the bottom of the ship, between its keel and the first strake or plank on either side. The sails exerted great pressure on these boards, and when they leaked they had to be attended to immediately. This seam normally had to be repaired from the outside, so the captain generally grounded the ship on an outgoing tide, careened it by pulling it over on its side, and waited for the seam to be uncovered. If it wasn't already loaded as a separate store, ships' carpenters could easily make pitch by igniting the tar in a kettle, or even in a pit on the shore, until the tar had thickened to the consistency of pitch. There was no time to waste. The tide would quickly rise leaving the job half done unless the pitch was ready for use. The expression "the devil to pay" is a shortened form of "the devil to pay [or patch] and no pitch hot." Ships' captains would not think kindly of tars who hadn't heated the pitch sufficiently when the devil seam was uncovered. "It was a maxim of Captain Swosser's," said Charles Dickens's Mrs. Badger in *Bleak House*, "speaking in his figurative naval manner, that when you make pitch hot, you cannot make it too hot."[13]

A similar expression, "between the devil and the deep blue sea," is often used whenever a person is faced with two disagreeable alternatives. It arose from a more urgent naval situation when, with no shore nearby, the devil seam had to be patched at sea. Then the ship was heeled on its side and a sailor lowered overboard with his caulking irons and bucket of pitch, facing the twin dangers of being drowned or crushed by the waves against the side of the ship.[14]

———

With tar and pitch so essential to the navies and merchant fleets of the day, all European countries sought a secure source of trees. With most of their conifer forests cut down long before, Europeans by the eighteenth century were dependent on forests of *Pinus sylvestris* that lined the Baltic Sea. Shipments of tar and pitch, and some small cargoes of turpentine spirits, were regularly sent to England from its colonies in New England, Virginia, and the Carolinas. Despite England's encouragement, however, these shipments were relatively small. As late as 1704, the colonies' tar and pitch represented only 872 barrels out of the 61,525 barrels that England imported. By 1718 colonial exports of tar and pitch to England had increased to 82,084 barrels, a factor of nearly a hundred. A few years later it was estimated that 120 vessels were engaged in the coastwise and transatlantic trade of colonial tar, pitch, and raw gum.[15]

This great upsurge occurred when Swedish control of naval stores shipments from the Baltic induced England to offer bounties to the colonies to make tar and pitch for the British navy. In 1705 a bounty of five to ten shillings per barrel of tar and pitch was established. With this inducement, the industry began to shift from New England to the vast southern pineries. Virginia and the Carolinas responded quickly, with South Carolina taking the lead, but after the bounties were reduced, South Carolina concentrated on indigo, rice, and cotton while the small farmers of its neighbor to the north maintained and increased their interest in tar and pitch. North Carolina would go on to produce 70 percent of the tar exported from all the American colonies; for 150 years, from 1720 to 1870, it would be the greatest producer of naval stores in the world.[16]

Though the naval stores industry was no doubt based on European models, the forms of the industry changed in response to American conditions, in particular, to the unique set of conditions offered by the Piney Woods. This is abundantly clear with regard to the making of tar, the process of which evolved in an eccentric direction in Virginia and North Carolina.

England preferred the colonists to make tar by the expert Swedish methods. The Swedes made tar from "green" or growing trees. They removed the bark from a growing tree up to a height of eight feet and let it stand for a year or two while the debarked wood became heavily saturated with resin. Then the tree was cut down and the debarked portion cut up and burned in the kiln. This procedure, known as the "East Country" method, yielded purer tar and more of it, but it required more work and more workers than the colonists were willing to give it. North Carolina tar burners made tar by collecting lightwood from the forest floor and burning it in their homemade kilns. This tar, known as "common" tar, had a poorer quality. Despite frequent entreaties and twice the bounty for green tar than for common tar, colonists continued to make it as it suited them.[17]

Hikers in some of North Carolina's coastal forests often run across the remains of tar kilns. The most visible clue is an unexplained depression in the soil, with perhaps some signs of charcoal just below the surface. I spent a few hours one day examining a tar kiln on the property of Aubrey Shaw in Bladen County. A retired high school biology teacher, now deceased, Shaw liked

show school kids and Scouts what tar burners did and how the turpen-
iers tapped trees and collected the raw gum. Shaw lived in Roseboro, but
..ᴄ owned about four hundred acres not far from there, in the nearby Lake
Creek Community, all that remained of family property that once amounted
to several thousand acres, most of it in longleaf pine.

Shaw met me at his little tin-roofed retreat with curved, heart cedar porch
rails that he called Teaberry Ridge. There is little longleaf left on his property
now, and most of that Shaw planted more than forty years ago. After crossing Sampson County from the interstate I turned onto a two-lane country
road that parallels the South River not far away. The river rises near the town
of Benson and runs southeast, joining the Black River somewhere below
the former railroad town of Ivanhoe. The tannin-stained waters of the Black
flow placidly across the gently sloping Coastal Plain and join the Cape Fear
River about forty miles above Wilmington. Teaberry Ridge stands in eastern Bladen County, its porch railing facing the South River and its back door
opening toward the Cape Fear — in 1860 the greatest artery of the worldwide
naval stores trade.

Shaw greeted me cheerily. He was seventy-four years old at the time, an
age he shared with the oldest longleaf pine trees on his land. On the unseasonably warm January day he was dressed in jeans, a plaid shirt, and a
blue ball cap. We talked awhile on the cabin porch. He said that his family
had been on this land since the 1700s. He was a seventh-generation North
Carolinian. "My great grandfather was George Washington Bannerman,"
he said. "I would say he owned a couple of thousand acres of land and a lot
of that was in longleaf pine. He was in turpentine heavy until about 1900.
But the trees are gone. I know where they were, and there's not but just a
handful of longleaf pine on it now. That gives you an idea of what there used
to be here. Sherman burned the old family house down during the Civil
War but my great uncle rebuilt it on another spot. My great granddaddy
was a chaplain in the Civil War. I remember him just a little bit. I could have
gotten a whole bunch of data from him."

He used the word "data" several times that day to mean information.
He was curious about the old ways, especially about turpentining and tar
making, and he yearned for better records. "Those people back then didn't
keep the data," he said. He was concerned about that and had questioned
a lot of his neighbors, asking them what they remembered about the processes. One of his neighbors, an African American named Tommy Johnson,
recalled how they used to make tar kilns. Shaw had sketched the different
stages of the process Johnson described in ten separate drawings. He kept

the sketches, dated July 3, 1994, in a business-sized envelope. In step five, instead of sketching the kiln as he had in the previous steps, he had written: "Placing the lightwood in kiln. Refer to No. 4 diagram—(am tired of drawing)."

How had he become interested in longleaf pine and naval stores? "I reckon it's best to put it like this," he said. "I grew up here and it's just inherited. It's something I'm close to." We listened to the sounds of hunting dogs yelping in the distance. "The older I get, the more I see the value of history. I didn't think about it at all for a long time. I do just enough to keep the old art living. Everyone else has passed away."

The two of us got into his truck and drove it along a sandy woodland path. We stopped at a shed not far from his demonstration area where he showed me two chunks of sawed pine about two inches square. One was of longleaf pine sapwood and the other was of heartwood. The heartwood felt dense and heavy. Then he showed me three pieces of gnarled, knotty wood. I thought they were rotten until I hefted one of them.

"That's what's called 'lighterd' or 'fat lightwood,'" he said. "In the forests of today, you don't find lightwood because the trees are normally cut before they grow old enough to form it." After the old trees died and fell, the sapwood rotted away, leaving the dense heartwood that was practically impervious to decay. Lightwood (sometimes called "lighterwood" or "fatwood") was the wood—from stumps, knots, and limbs—that had become especially saturated with resin. When people first settled in the region, lightwood from branches and dead trees littered the forest floor. According to one nineteenth-century writer, "With these dry bones the surface is very much covered—not entirely covered, but so thick as to fill the body of a horse-cart, in many places, in a square of ten yards, and every so often in a square of ten feet." Because they were saturated with flammable resins, lightwood pieces burned easily; the settlers used them as torches and burned them in tar kilns to make tar and pitch.[18]

We stopped at one of the tar kilns Shaw had discovered and excavated. The kiln was about twenty feet long, rectangularly shaped, and sloping toward one end. It was covered with oak leaves and pine straw. We brushed off the debris to uncover the charred wood remains from the last tar burning and the blackened bed of the kiln. Its rectangular shape was unfamiliar to me. Most of the tar kilns I had seen and read about had been oval-shaped, about twelve to fourteen inches deep with a diameter of anywhere from fourteen to thirty feet. Tar burners gathered lightwood, split it into kindling about two to three feet long and two to four inches wide, and stacked it perhaps fourteen feet high. The finished stack looked like an inverted cone, with the wood sloping toward the center. A gutter at the bottom of the depression, sometimes with a hollow log in it, conveyed the hot, liquid tar to the tar hole about six feet from the circumference of the kiln. Careful tar burners would also line the kiln floor with clay.[19]

At this point the tar burners covered the kindling with pine straw, freshly cut pine branches, and sod, making sure it was airtight. They forced burning lightwood down through an opening on top of the kiln until it blazed up, and then they dampened the blaze by covering or partly covering the opening. The trick was to get the wood to burn slowly so that the tar was sweated out of it. It was a ticklish process. If the burners let too much air into the stack, the tar could be spoiled; too little air could cause an explosion that killed or scalded the tar burners. The tar burners tended the kiln day and night for a week or even more, keeping the fire burning steadily until the flow of tar ceased. To many travelers through the Piney Woods, tar burners seemed especially sinister. "So dense [was this dark wilderness of pines] . . . that it seemed as if we had entered a realm of sighing and moaning," wrote one visitor near the Cape Fear River in 1853. "The somber appearance which is given to the country by its extensive pine forests is by no means cheered by the tar kilns which meet the eye here and there as you pass . . . and which resemble burning volcanoes on a small scale . . . surrounded by an unearthly set of black figures in human shape, thrusting long pikes into the agonizing structure."[20]

On the second day the tar began to run, gradually collecting at the bottom of the kiln. To prevent air from passing into the middle of the kiln and igniting the wood, the workers blocked the hollow pipe for about thirty-six hours; then the passage was unblocked and the tar was allowed to run into the hole. Enormous amounts of wood were required to make tar. A kiln thirty feet in diameter accommodated about 180 cords of wood, which produced about 180 thirty-two-gallon barrels of tar. Timothy Silver calculates that North Carolina's total exports of tar and pitch in 1753 required 75,000 cords of wood, or a stack four feet high, four feet long, and 113 miles long. The tar itself, a thick black substance, was basically the resin of the tree along with a mixture of hydrocarbons, alcohols, and other compounds.[21]

Over the next few days the kiln settled as the wood burned. When it had stopped burning, it might contain a mound of charcoal about five feet high, a somewhat valuable by-product especially favored by blacksmiths. A kiln producing 150 barrels of tar might make 800 bushels of charcoal.[22]

In addition to the rectangular tar kiln, Shaw found several oval tar kilns on his property. In fact, there were scores of kilns all over the woods, he told me, evidence of how commonly tar was produced.

"Look here," he said. "The bottom of this kiln is hard as a brick. I would think that they would use this kiln over and over, don't you? Once you built one, you kept it. You know, you couldn't just put a tar kiln anywhere. You'd have to dig down to the hardpan layer. If you dug one on sandy soil, you'd lose a lot of tar. You didn't want to blaze it. Just sweat it. I've looked at these all my life and I used to think, well, you know, you just dig a hole. No sir. They knew what they were doing. It was a skill. I'm just beginning to learn that."

After we returned to his cabin, Shaw showed me another set of notes he had made from a conversation he had had with Ernest Bannerman, another African American neighbor:

As told by Ernest Bannerman
As things were 1930
Bladen County—Lake Creek Community—
"Tarkull"—Tar Kiln

To make and burn a kill.
No. People—2
To dig—2 days
Lightwood—40 cart loads—12 days to cut & haul
To cover—1 day
To burn—7 days
Tar produced—12 50 gal. barrels
Hauling—50 cents/bbl.
Rent for lightwood—?
Earnings—Per man. About $1.25/day.

I asked Shaw what people did with the tar they made. In the old days, he said, they brought it over to the Black River and rafted it to Wilmington to sell. "People would use a little of it on the farm to make pitch. You'd take the raw tar and boil it some more and it got more sticky, more compact. That would be pitch. We'd use it for patching boats and after castrating hogs you'd swab some tar on them. We didn't use too much, but it was sold."

While we sat on his porch and ate our lunches, Shaw told me his own explanation of the origin of the term "Tar Heel." While ladling the tar into barrels, the burners inevitably spilled some onto the ground and it stuck to their bare feet, picking up a lot of pine straw which would be permanently attached until the tar wore off.

The story had the ring of experiential truth about it and would explain the derogatory origins of the term. According to the "official" version, "Tar Heel" was first an epithet and then a badge of honor during the Civil War. After one battle in which the North Carolinians failed to hold a hill, a group of Mississippi soldiers told them that they had forgotten to tar their heels that morning. But on another occasion, it was the Virginians who had re-treated and a smart-alecky North Carolinian who told them that Jefferson Davis was going to put tar on their heels "to make them stick better in the next fight." The sight of tar burners walking about the woods with tufts of pine straw sticking to their shoes may have been just comical enough to make "Tar Heel" a natural expression.[23]

In between bites of his cheddar and slugs of his grape drink, Shaw told me how he'd like to burn a tar kiln someday and let kids watch it. "I think that would be fascinating for them. We could build a little lean-to like they used to do and bring some meat and sweet potatoes out with us. The kids could camp out, roast the potatoes, stay the night. Now, that would stick with them, wouldn't it?"

SEVEN

Getting Turpentine

The making of turpentine and tar is almost the sole
business of the thinly settled population of the pine lands.
—Edmund Ruffin, "Notes of a Steam Journey" (1840)

It's hard for us to imagine today how bewitching the word "turpentine" once was, how powerful its allure. "Getting turpentine" became a mania in the nineteenth century, driving thousands into the forests of longleaf pine where they fought off the heat, rattlesnakes, ticks, chiggers, and loneliness while cutting "boxes" into the living trees, chipping trees, spooning the raw gum into buckets, and carting heavy barrels of it along woodland pathways to the turpentine stills. Travel in the southern pinelands of the mid-nineteenth century was a full sensory experience starting with the resinous fragrance of the pines themselves as well as the whiff of turpentine from the stills scattered throughout the woods. Tar kilns smoldered by the roadside, sending up a pall of smoke. Chippers and dippers moved quickly from tree to tree and shouted their tallies. The eerie sight of boxed trees slicked white with sticky gum never failed to impress travelers on the lonely roads through the pine woods: "Our way lay through an unbroken forest and as the wind swept fiercely through it, the tall dark pines which towered on either side, moaned and sighed like a legion of unhappy spirits let loose from the dark abodes below. Occasionally we came upon a patch of woods where the turpentine-gatherer had been at work, and the white faces of the 'tapped' trees, gleaming through the darkness, seemed an army of 'sheeted ghosts' closing steadily around us."[1]

Hundreds of barrels of gum lay beside the road, waiting to be hauled away on ox-drawn carts to a still or a railroad stop. Thousands of stills turned countless barrels of gum into sometimes (but not always) profitable barrels of turpentine spirits and rosin. Communities formed at the river landings where white men and black men awaited the winter freshets along the Blackwater and Nottoway Rivers, the Tar and Cape Fear, the Waccamaw, PeeDee, Sampit, and Congaree; the Savannah, Oconee, Altamaha, and Chattahoochee; the St. Johns and Appalachicola; the Conecuh and the Alabama; the Pearl and the Pascagoula; the Calcasieu and the Red; the Trinity and the Sabine. On their high waters barrels of naval stores floated on rafts, poleboats, flatboats, pirogues, and stern-wheelers to New Orleans, Mobile, Pensacola, Jacksonville, Savannah, Charleston, and Wilmington, where so much traffic congregated at one time that you could walk from one side of the river to the other on the backs of scows waiting to unload. On the wharves, barrels towered three high, tens of thousands of them, stretching for acres. On hotel verandas commission merchants, wholesalers, and retailers knocked back whiskeys after complex financial dealings that sent the barrels rolling onto three- and four-masted schooners, lowered by winches into their holds to ride the coast-wise journey to New York, Philadelphia, Boston, or the longer passage to London, Bordeaux, Antwerp, and other transatlantic ports.

The longleaf pine forests were the greatest turpentine forests in the country. They helped make the Southeast a worldwide economic player, and they made the state of North Carolina a naval stores kingpin. By 1840 North Carolina had a near-monopoly on the production of turpentine and rosin in the United States; its ports shipped nearly 96 percent of the total. In 1850 there were 785 stills in the state, more than ten times the number of stills in the other southern states combined. By 1860 the total value of the industry was almost $12 million and of that, North Carolina produced the lion's share.[2]

It was a vital and colorful industry, yet a wasteful and destructive one, too. Turpentining exploited the labor of tens of thousands of workers, made some men rich, created even more paupers, and contributed mightily to the devastation of the longleaf pine forest.

━━━━━━━━━

Turpentining produced two naval stores products — spirits of turpentine and rosin — and it was a lengthy and somewhat complicated process. Unlike tar, which was made from the wood of a dead tree or limb, turpentine and rosin were the distilled products of the living tree. Like tar and pitch, farmers could produce turpentine and rosin for their own uses, or they could produce it for the market. In the eighteenth and nineteenth centuries turpentining developed into a major industry, first in North Carolina and then elsewhere in the South, and employed mostly slaves who performed specific tasks in a highly standardized way. These turpentine operations were based on principles of organized work, repetitive tasks, and worker castes.[3]

Turpentining was a nearly year-round operation that began in winter with "boxing" the pines—chopping a pocket deep into the base of the tree. Beginning in March and running through October, "chippers" roamed the forest cutting long, shallow streaks immediately above the box with a sharp tool called a "hack." The hack cut through the bark and into the wood of the tree, causing resin to run from the wounds down the streak channel and collect in the "box" below. Each week, the chippers hacked two fresh streaks above the previous streaks in the form of a "V." As the turpentine season went on, the streaks rose above the box, creating an open area of resin-covered wood known as the "face." Depending on the width of the chip, faces could rise a foot or two each year.

Once every week or two, a "dipper" ladled the gum that had collected in the boxes into a barrel. Ox carts drew the barrels of gum to the turpentine still, where the "stiller" distilled them into spirits of turpentine and rosin. Keeping the whole operation going were the coopers, who made the barrels in which the turpentine and rosin were shipped.

The gummy pine resin that rushes to the site of the wound is not its sap. Sap is the name for the nutrients and water that move up from the tree's roots to its branches and needles through recently produced hollow cells, called the xylem or sapwood. Pine resin is an oleoresin, a soft gum made in specialized cells within the tree and carried in vertical ducts that run up through the sapwood and connect to horizontal canals that run out to the bark. Its main function is to protect the tree. When a pine tree is wounded, the gum flows to the site, covering the wound and protecting it from insect invasion. When a turpentine worker hacked through the bark of the tree, he cut through its cambium and sliced off the ends of the resin canals, causing the resin to run. Eventually the gum hardened as the volatile elements evaporated, which is why the chipper returned each week to cut another streak and stimulate a fresh discharge. The resin flows better during warm weather and thus in the southern states the turpentine-collecting season—the time when the chipping and dipping activities were carried out—traditionally began in early spring and ended in the fall.[4]

The remarkable thing about pine gum is that a wound stimulates the tree to produce new resin-carrying canals. Experiments early in the twentieth century showed that almost 90 percent of the resin flow occurs during the first three days after chipping. Thereafter, the resin ducts fill with hardened resin and the flow is stopped. But the resin cells immediately above the chip are filling, so that when the chipper made another wound a week later he caused the new resin to flow, at the same time creating new cells for another supply of resin which would be tapped the following week. It might be better to think of the pine tree as a gum-making factory or mine rather than a gum reservoir with a finite and exhaustible supply.[5]

The oleoresin of the longleaf pine tree consists not only of resin but also of an "essential oil." Both of these products are valuable and are separated during the distillation process into the liquid turpentine (sometimes called spirits of turpentine) and rosin, a liquid at first but hardening into a solid. The oil is volatile, meaning that it is easily vaporized. On a hot day it's the resin transpiring through pine needles that is most responsible for the fresh fragrance of a pine forest. This fragrance has no known use, but it is responsible for the ancient attraction of pine forests and the belief in their health benefits.[6]

Whereas tar could be produced by a few workers over a few weeks, industrial turpentining required the efficient coordination of many different workers throughout the year in cutting and cornering boxes, chipping the streaks, dipping the raw gum, carting it to the distillery, making barrels, distilling, raking pine straw away from the trees, and gathering wood. Turpentine work was grueling, hot work. Slaves were thought to prefer it to working in the fields, but not always. Sold to a turpentine operation in Brunswick County, North Carolina, in the 1850s, a slave named Ned escaped and returned to Virginia where he was caught. The white man who found him wrote to the slave's owner: "The work and the manner of life in making turpentine he cannot stand. It is hard work and would kill him by piecemeal, and he had rather be killed at once."[7]

How important the turpentine industry was can be judged by the number of ways that turpentine and rosin were used.

Turpentine's applications stretch back into antiquity. The Romans used it as an antidote to depression and apoplexy. Later, spirits of turpentine joined epsom salts, liniments, emetics, tourniquets, syringes, splints, amputating and trepanning instruments, and other medical supplies that were loaded aboard ships before each voyage. During naval battles, surgeons injected hot turpentine directly into wounds and coated the stumps of amputated limbs with it, and as late as World War I, battlefield doctors treated hemorrhages from bullet wounds with turpentine. An intimidating number of other medical conditions were treated with turpentine, including epilepsy, tetanus, convulsions, diabetes, hysteria, typhus, yellow fever, dysentery, and even madness. It was widely used as a remedy for a variety of pulmonary illnesses, including tuberculosis.[8]

Turpentine spirits, like rosin, had many applications around the home and the farm. Like other gum derivatives, it long had a folk reputation, whether medically sanctioned or not, as an all-purpose remedy. For tapeworm, American physicians in the nineteenth century recommended a large dose of turpentine taken internally followed by a spoonful or two of castor oil. Have a cut, scratch, or nail puncture wound? Turpentine on a clean rag was an antiseptic. Bronchitis? Inhale the steam of turpentine mixed with water. Rheumatism? Mix one cup each of apple vinegar, turpentine, kerosene, and whiskey (and throw in a nickel's worth of gum of camphor).[9]

Blisters, burns, corns? Lumbago, sciatica, neuralgia, pleurisy? Gums hurting from an abscessed tooth? Insect bites? Turpentine was applied locally for relief of all of these conditions.

Turpentine cleaned carpets (three tablespoons of turpentine added to a quart of water and applied with a sponge) and removed paint stains and tar from clothing (with a little ammonia). A few drops in a clothes closet or chest kept away moths; in mouseholes, it chased away mice. Turpentine and beeswax made a fine floor polish; a mixture of turpentine and "sweet oil" polished furniture. Confederate soldiers waterproofed their leather boots with a mélange of tallow wax (half a pound), hog's lard (four ounces), turpentine (two ounces), and beeswax (two ounces). Candlewicks burned brighter when they were first dipped in turpentine.[10]

New industrial uses for spirits of turpentine as a solvent in the manufacture of rubber and as an illuminant were discovered in the nineteenth century, and created a demand for turpentine that began to rival that for tar and pitch. Mixed with castor oil or with alcohol (a mixture known as camphene) and burned in lamps, turpentine provided bright light with little odor. Other industrial uses included the manufacture of oil colors, varnishes, and paints.

As useful and desirable as turpentine was, however, it was rosin that gradually commanded the attention of manufacturers in the nineteenth century. It had always been an essential product on farms and in households. Farmers used rosin as a preservative, a waterproofing agent on boots, an ingredient in shoe polish, and in making sealing wax for preserving fruit. Cooks sometimes heated a pot full of rosin and cooked potatoes in it. During the hog-killing season, rosin added to a vat of boiling water coated the hog's skin, making it easier to scrape off the hair. Farmers could do the same to remove the feathers of geese and ducks.

Early manufacturers used it mainly for waterproofing and soap making; even in the early twentieth century, a jellylike soap made from surplus fatty meat and lye was hardened by the addition of a little pine rosin. But as rosin grades improved later in the century, more uses were found for each of the grades. In general, buyers required the lightest grades, graded as "Pale" or white rosins ("Waterwhite," "Windowglass," and "N"), for the making of fine soaps, varnishes, candles, India ink, ointments, and in lithography and paper sizing. The darker and cheaper medium grades ("M," "K," "I") were ingredients in the making of yellow soap, sealing wax, and pharmaceutical products, and for other purposes. The darkest common grades ("H," "G," "F," "E," "D," and "B") were used in the making of brewer's pitch (used to coat the inside of beer barrels) and for distilling rosin oil, which was useful in the manufacture of lubricants. The discriminating and thrifty buyer used the appropriate grade of rosin for the appropriate purpose.[11]

The most important worker in a turpentine operation was the distiller. A distiller often made all the difference between producing a profit or a loss for his owner or employer by raising or lowering the grade of the turpentine and rosin. A careless, inexperienced, or drunk stiller could even cause an explosion, maiming or killing himself and others nearby, as well as burning the community or even the city to the ground. This happened often enough that experienced slave distillers were among the most valuable workers in the industry.

Turpentine had been difficult to distill before the 1830s. The iron retorts used until then were heavy and expensive. Thus stills almost always were concentrated in major cities, and the raw gum from the backwoods was rafted or boated directly there for distillation or even shipped farther, to the buyer in New York, Boston, or London, who would distill it himself. Sending it out of state for processing was a distinct loss of value for North Carolina commerce. In 1834 Scotch-Irish liquor makers introduced the copper still to North Carolina, a technological advance that was one of the few genuine revolutions the industry ever achieved.[12]

The new copper still was relatively cheap and light. It made better turpentine than the iron ones and was more easily cleaned. Better yet, it was movable. The distillery could easily be erected in the backwoods, dismantled, and set up somewhere else. This mobility fueled the spectacular advance of the turpentine industry throughout the Southeast from the 1840s on.

One May I drove from Raleigh to the Georgia Agrirama in Tifton to see spirits of turpentine and rosin made in an old-fashioned copper still. The Georgia Agrirama is a living museum, with some forty buildings collected from throughout the state and interpreters in period dress who depict the styles of agricultural, industrial, and commercial life in Georgia of the 1870s to 1900. The dusty 95-acre tract just off I-75 is as authentic a look at nineteenth-century Piney Woods life and vernacular architecture as you'll find. Twenty-five of the buildings are original structures, including a farmhouse, one-room school, country church, water-powered gristmill, steam-driven sawmill, and turpentine distillery. Each May, the Agrirama fires up its old turpentine still and demonstrates the ancient art of turning raw pine gum into usable products.

After spending the night in Macon, I drove south early the next morning and reached Tifton in about two hours. I wanted to get there early because I had been permitted to photograph the process and interview the workers. There was a holiday feeling to the occasion. Most of the attendees were middle-aged or elderly men, nearly all of them white. Many old-timers in the area who were once involved with the turpentine industry eagerly await the firing of the still each year and enjoy seeing it in action one more time. What used to be the daily life of these elderly men is now a historical curiosity, and they enjoy talking to tourists about the way they made turpentine in the old days.

The Agrirama's turpentine still is a two-story open shed with a brick chimney running from the ground floor to the tin-covered, gabled roof. The main structural supports are wooden poles set at the corners. On the ground floor is a large brick furnace. Visible on the second floor is the "still cap," a straight-sided copper structure with a long goosenecked top. Below the still cap, and hidden inside the brick structure, is the kettle, and beneath that is the firebox where wood is loaded and burned to fuel the distillation process. The kettle contains the raw gum to be cooked. The vapor rises from the kettle to the still cap above and through the gooseneck into a coiled pipe called a "worm" that is contained inside a wooden water tower rising from the ground to the level of the second story. The vapor passing through the worm condenses into liquid form. Leading up to the second floor is the gum ramp with pole rails that guided the heavy, fifty-gallon barrels of gum pushed by the workers. Often these ramps were connected to a loading deck about twenty-five feet away that was about the same height as the wagon beds, which made it easier for workers to unload the heavy barrels of gum.

Buster Cole was the main interpreter for the event that day. He was a tall, lean black man who looked to be about sixty years old but could have been seventy. He was wearing a black cowboy hat, creamy white shirt, and tan pants. Cole had a singsong voice and described turpentine making with an entertaining, carnival type of patter. I climbed up to the second story of the still with him, and we examined several barrels of pine gum that had been trundled up the gum ramp and were being readied to charge the still. Each of the fifty-gallon barrels weighed about 420 pounds. The copper cap of the still had not yet been removed.

It was mid-morning and the yard was beginning to fill with people, their video cameras at the ready. A variety of interpreters bustled here and there about the still, the sleeves to their wide-cut period shirts flapping in the breeze. Firing a still takes about three hours and there are only a few dramatic moments, so during the time that it took to complete the process the crowd formed and reformed, new faces appearing and then disappearing, looking on for a few moments, eavesdropping on the garrulous old turpentiners, and asking questions.

Finally Cole began his spiel:

"This morning we're going to do a demonstration of turpentine stilling, which is a lost art. This here is an old-fashioned nine-barrel still. Back when I was a boy, there were seven-barrel, nine-barrel, and thirteen-barrel stills. We've got four barrels of crude gum we're going to work with today — we're not going to load up all nine barrels. Too expensive. When I was growing up, we used to sell a barrel of gum for three-ninety or four dollars a barrel. Now a barrel of crude sap runs anywhere from eighty-five to one hundred and twenty dollars a barrel. Back then, they'd pour in nine barrels of crude sap and cook it. They'd heat it from two hundred and ninety to three hundred and twenty degrees. They'd let the steam go up through the copper coil into the water tank, and that would condense the steam into a liquid and it would come out turpentine and water. The turpentine would float to the top because it's lighter than the water. They'd cook all the turpentine out and put it in barrels. What they had left was the rosin. They'd dip that out, put it in barrels and ship it to different places for different things. Out of nine barrels of resin you might get two barrels of turpentine and about six-and-a-half barrels of rosin. That's the average.

"When it's brought from the tree, it's called "sap," "draw" or "resin" — R-E-S-I-N. [He spelled the word slowly.] After it's distilled, the by-product is called rosin — R-O-S-I-N.

"There are one hundred and twenty-nine uses for turpentine and rosin combined. You could use it for gunpowder, glass, fiberglass, acrylic, polyester, chewing gum, costume jewelry, tile, Elmer's Glue, shellac, soap, shoe polish, soldering compound, rubber, dry-cell batteries. In gymnastics you had a rosin bag, in bowling you had a rosin bag, tennis players got a rosin bag, baseball pitcher had a rosin bag, too. A violin player put rosin on his bow, a dancer put it on his shoes, and they made nine cosmetics out of it. Three perfumes were made out of turpentine, twenty-three percent of all your extracts got turpentine in it, pine solvent got turpentine in it, pine spray got turpentine in it, disinfectant got turpentine in it. They make paint remover and paint thinner, and finger nail polish got turpentine in it."

The crowd grinned at Cole's animated talk.

"There are one hundred and twenty-nine uses for these two products combined and rosin is bought and sold by the grades. That's right, you buy it by the grades. There's twelve grades of rosin. The best grade is 'X'; that's excellent. 'WW,' that's water white. 'WG,' that's window glass. And then there's Nancy. Mary. Katie. Isaac. Harry. George. Frank. Dolly. And Betty. Nancy is the lightest and Betty is the darkest. They were slave names. See, they'd take a piece of the rosin and look at it and say, 'Why that's about the color of Nancy there.' Another run of rosin and they'd say, 'That's about the color of ol' George. And this is the color of Betty.'"

Cole and another black man mounted the stairs to the second floor of the still where they would "charge" the still with four barrels of gum. They removed the copper cap to the still, revealing a black hole below, and wrestled one of the 420-pound barrels to the opening where they upended it. Slowly the thick, whitish gum oozed out, pocked with all kinds of trash—bark chips, pine needles, leaves, small pieces of wood. It all dropped into the kettle below like huge dollops of cookie dough. Each of the four barrels was loaded in a similar way.

At this point the stiller and his assistant—two white men—materialized and fitted the cap over the still and its gooseneck onto the piece of copper coil that emerged from the water tank. They sealed both joints tightly using a wadding of cheesecloth covered in plaster of Paris. One hundred years ago, stillers would have used clay as the sealant. It was crucial to make a good seal. Escaped vapor not only represented the loss of product but also posed the more immediate threat of fire. A distillery fire was a life-threatening event about which you could do little, according to Buster Cole, except run for your life.

Fires happened all the time around distilleries despite the precautions workers took. Workers spilled flammable resin on the decking, and some resin fell to the ground floor near the firebox. The vapor from the distillation process was about as flammable as the gum. In Wilmington, twenty distillery fires occurred from 1842 to 1853, and many fires destroyed wharves and other places where turpentine was stored. Turpentine fires sometimes incinerated an entire community. Anyone who ran a still was living a dangerous life and posed a threat to the community.[13]

Once the still was sealed, the stiller lit the fire in the firebox below. The crowd had drifted away. The logging train was running now, whistling ferociously every time it carried a bunch of visitors into the woods at the end of the run and back. The steam-powered sawmill was sending up an intermittent scream as the millers competed for visitors with the distillers nearby.

Black smoke drifted out of the chimney above the still. Distillation separates the constituent parts of any substance. Pine resin is distilled into two distinct forms: turpentine (the liquid oil, or spirits) and rosin. The two products are separated because they each have a different evaporation point. As the gum in the kettle begins to heat up, it vaporizes between 212 and 316 degrees Fahrenheit into a mix of water and turpentine. Rosin's boiling point is much higher, about 392 degrees, but the stiller makes sure that the temperature never reaches this level because it would spoil the rosin and contaminate the turpentine. After all the turpentine is vaporized from the gum, the hot liquid rosin is discharged from the kettle.[14]

Much of the stiller's skill lay in knowing how to manipulate the fire. It had to be hot enough to vaporize the turpentine and water, but not so hot that the gum boiled. If it did, both turpentine and rosin could be spoiled and the still could explode. At various junctures the stiller added water through a funnel in the still cap or pulled the wood from the fire, both of which reduced the amount of heat. He could also add wood and raise the temperature if needed. At the Agrirama, the stiller had a new-fangled electronic gauge to keep him informed of the temperature, but in the old days he had to listen to the sounds of the cooking, a process known as "sounding the still."

I spent much of the ensuing two hours watching as the stiller checked the temperature gauge, adding water and pulling the wood out of the furnace, then building the fire back up again. Every now and then, for effect, perhaps, or for the cameras belonging to those of us who were still watching, he would "sound" his still. Where the outlet pipe emerged from the tank, he put his ear close and listened intently. I asked him what he was listening for.

"The higher the pitch, the hotter the gum," he said. "The lower the pitch, the cooler it is. At first it sounds like 'Whoa-whoa-whoa-whoa.' Now it's 'Wha-wha-wha-wha.' If it gets too hot it'll sound like 'Wa-wa-wa-wa-wa-wa-wa.' It's kind of like a frying sound. You ever hear grease frying? When it gets hotter and hotter it's a-really popping. That's what you're listening for. C'mon over here and listen." I put my ear to the pipe and heard a distant roaring sound, a metallic crackling like the static behind an old 78-rpm recording .

After the gum cooked awhile, the first distillate of water and turpentine began to flow out of a pipe on the far side of the water tank and into the first of three separator barrels. Turpentine spirits are lighter than water, so they rise to the top of the barrel. There they were drawn off by another pipe into a second barrel. In the second barrel, turpentine and whatever water was still present were separated again, and the turpentine spirits were carried to a third barrel in a similar way. When the third barrel was full of pure turpentine, the bung hole was plugged and the barrel hauled away. Buster Cole held a bottle under the flow of liquid coming from the water tank until it was half full and held it up for his small audience. The turpentine layer, a green, coppery-colored liquid, floated cloudily on top of the water layer.

"There's your turpentine and water," Cole said.

"Oh, yeah," said an elderly man. "That sure smells good."

The stiller walked over several times to take a sample of the water-turpentine mix as it emerged from the condensing coils, inspecting the proportions of both. At first, the layers were about equal, but as the distillation process continued over the next two hours and the turpentine was gradually worked out of the gum, the spirits level rose higher and narrower until it was merely a thin sliver riding atop a thick water wedge. That was the stiller's signal that the charge was just about finished. The temperature had been rising all along, but now it was allowed to rise to about 315 degrees Fahrenheit to keep the rosin hot; then the stiller pulled the wood from the fire and prepared to discharge the rosin.

If there was any drama in the entire distillation process, it occurred when the hot rosin came shooting out of the still via a small tailgate at the opposite side of the firebox. Whereas it took turpentine a couple of hours to be distilled, the rosin was discharged in an instant of steam and shouting into a stack of three wooden frames about seven feet long, each one of which had been lined with a different gauge of filter. The topmost filter was made of a coarse metal screen that trapped pieces of bark, chips, pinecones, and other trash present in the gum. Below it a second filter consisting of a finer mesh caught smaller debris; at the bottom was the third filter, which was lined with a cotton batting that strained the finest trash. A wooden tray at the very bottom collected the steaming, filtered rosin.

When the discharge was completed, Buster and another worker lifted the topmost filter and walked about twenty paces away, emptying the dross onto the ground and raking it, another safety precaution. The chips were once sold as a fire-starter. The two black men ladled the hot liquid into barrels using buckets attached to long handles. The rosin was a translucent maple-syrup brown; it would be allowed to cool and solidify over the next forty-eight hours.

When Cole had finished his work, I asked him what grade of rosin the gum had made. "This is going to be about an Isaac," he said. "I could tell that even before the rosin came out."

"You mean you could look at the gum itself and tell what grade rosin it was going to make?"

"That's right. I know all the grades. This is my life's work."

"How did you get into the turpentine business?" I asked.

"Well, my Dad owned a turpentine still like this one," he said. "He and my uncle, they had to run it in a white man's name. They ran it for thirty-one years. Basically they sold it to him and my Dad ran it. We turpentined right here in Tift County, Georgia. We moved to Tifton in 1926 and I started working making barrels when I was nine years old. I'd make twenty-one barrels every evening after school and got five cents a barrel. Which was a lot of money in 1926. And then I'd work in the woods in the summer time and around the turpentine still and make my summer money. I graduated at fifteen years old. I went to college for one year; I was going to take up law. But see at that time it was unheard of for a black man to put up an office in the South. So I told Daddy I didn't want to go back to college. So he told me, 'Well, what are you going to do?' I told him I was going to work with him until I made up my mind. It was thirty-one years before I made up my mind.

"I enjoyed it. You know, when I was a boy growing up, people weren't afraid of work. People now want jobs with all the work taken out of it." Buster had a small audience now of mostly gray-haired men, and they laughed appreciatively. "I been retired about eleven years now. I didn't know it was so hard to do nothing until I retired. I ain't learned to do it yet, but I'm working on it. I'm working on it. One of these days I'll learn how to do nothing."

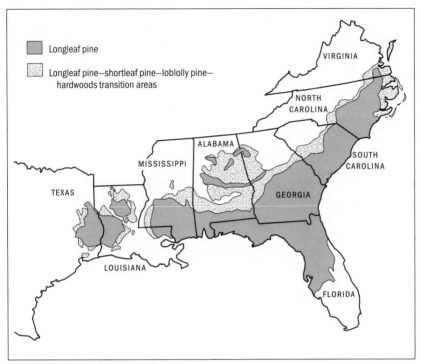

Longleaf pine

Longleaf pine—shortleaf pine—loblolly pine—hardwoods transition areas

VIRGINIA

NORTH CAROLINA

ALABAMA

MISSISSIPPI

SOUTH CAROLINA

TEXAS

GEORGIA

LOUISIANA

FLORIDA

The longleaf pine range (Adapted courtesy Cecil Frost and Tall Timbers Research Station, Tallahassee)

Reportedly based on a sketch by settler Peter Gordon, this 1734 engraving shows the year-old colony of Savannah, Georgia, pushing out into the surrounding pine forest. (Courtesy Library of Congress)

The Gulf Coast forests of longleaf pine often grew densely in the loamy soil. This virgin forest was photographed (date unknown) on the northern shore of Lake Pontchartrain, Louisiana. (Courtesy Forest History Society)

A virgin stand of longleaf pine in the East Texas Piney Woods region, photographed about 1910. (Courtesy Stephen F. Austin University)

This virgin longleaf stand was photographed (date unknown) on land belonging
to the Louisiana Central Lumber Company, near Clarks, Louisiana.
(Courtesy Forest History Society)

The Big Island Savanna in The Nature Conservancy's Green Swamp Preserve in North Carolina is typical of the thinly treed longleaf pine savannas once common in the Southeast. (Courtesy Cecil Frost, photographer)

Longleaf pine flourishes in North Carolina's Sandhills Game Land along with a healthy growth of wiregrass *(Aristida stricta)*. Wiregrass is a major fuel for the frequent fires that burned in much of the longleaf pine range. In the Gulf coastal states, wiregrass is replaced by bluestem grasses.
(Courtesy Cecil Frost, photographer)

The mesic or moist soils of this longleaf pine forest at Camp Shelby
in Mississippi host a high diversity of vegetation.
(Courtesy Bruce Sorrie, photographer)

Frequent fires in longleaf pine forests burned low to the ground. The
longleaf's thick bark protected it from all but the most intense fires.
(Lawrence S. Earley, photographer)

In the longleaf pine ecosystem, grasses and the plentiful pine needles
provided the fuel necessary to carry the life-giving fires.
(Lawrence S. Earley, photographer)

The illustration depicts various stages in the life of a longleaf pine. From the seed (life size, at left) springs the grass stage (inset), which gradually puts down a deep taproot and lateral roots. After several years the seedling begins a growth spurt to the bottlebrush stage (left), then begins to branch out with its "candles." A mature tree stands on the left, while an old-growth tree with its telltale flat top is at the far right. (Courtesy Anne Runyon)

Longleaf pine seeds are larger and heavier than the seeds of other southern pines. (Lawrence S. Earley, photographer)

The gopher tortoise is an endangered species that builds extensive burrows in the sandy soils of its habitat. (Lawrence S. Earley, photographer)

The red-cockaded woodpecker is the only woodpecker that excavates its cavity in living trees. (Courtesy Derrick Hamrick, photographer)

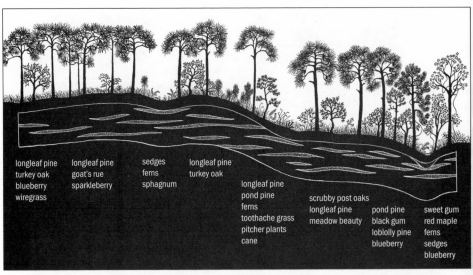

longleaf pine
turkey oak
blueberry
wiregrass

longleaf pine
goat's rue
sparkleberry

sedges
ferns
sphagnum

longleaf pine
turkey oak

longleaf pine
pond pine
ferns
toothache grass
pitcher plants
cane

scrubby post oaks
longleaf pine
meadow beauty

pond pine
black gum
loblolly pine
blueberry

sweet gum
red maple
ferns
sedges
blueberry

A sandhills longleaf pine community. Longleaf pine occurred in several different plant communities, including the sandhills longleaf association. The deep sands of the sandhills kept plant diversity low. (Courtesy Anne Runyon)

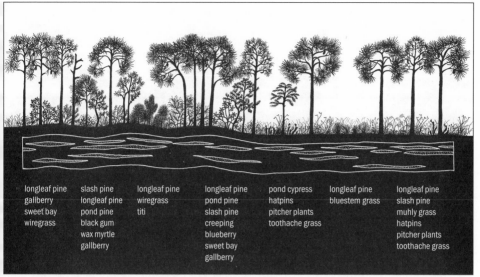

longleaf pine	slash pine	longleaf pine	longleaf pine	pond cypress	longleaf pine	longleaf pine
gallberry	longleaf pine	wiregrass	pond pine	hatpins	bluestem grass	slash pine
sweet bay	pond pine	titi	slash pine	pitcher plants		muhly grass
wiregrass	black gum		creeping	toothache grass		hatpins
	wax myrtle		blueberry			pitcher plants
	gallberry		sweet bay			toothache grass
			gallberry			

A flatwoods and savanna longleaf pine community. These longleaf sites have wet soils that host a much higher plant diversity than sandhills. Ecologists have found extraordinary levels of diversity in savannas. (Courtesy Anne Runyon)

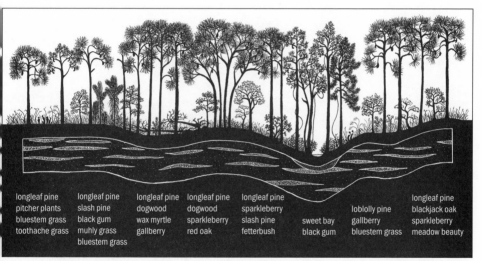

longleaf pine	longleaf pine	longleaf pine	longleaf pine	longleaf pine		loblolly pine	longleaf pine
pitcher plants	slash pine	dogwood	dogwood	sparkleberry		gallberry	blackjack oak
bluestem grass	black gum	wax myrtle	sparkleberry	slash pine	sweet bay	bluestem grass	sparkleberry
toothache grass	muhly grass	gallberry	red oak	fetterbush	black gum		meadow beauty
	bluestem grass						

A rolling hills longleaf community. On the loamy soils of the Gulf Coastal Plain in Alabama, Mississippi, and Louisiana, dense longleaf forests produced the largest trees and best timber. (Courtesy Anne Runyon)

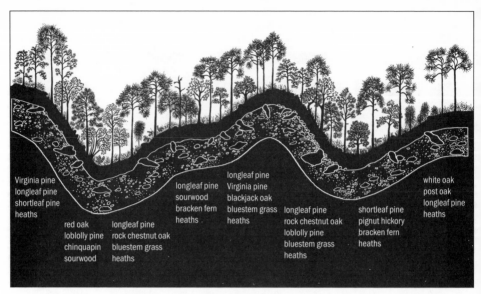

Virginia pine
longleaf pine
shortleaf pine
heaths

red oak
loblolly pine
chinquapin
sourwood

longleaf pine
rock chestnut oak
bluestem grass
heaths

longleaf pine
sourwood
bracken fern
heaths

longleaf pine
Virginia pine
blackjack oak
bluestem grass
heaths

longleaf pine
rock chestnut oak
loblolly pine
bluestem grass
heaths

shortleaf pine
pignut hickory
bracken fern
heaths

white oak
post oak
longleaf pine
heaths

A mountain longleaf community. Best known along the Coastal Plain, longleaf pine also grew in the higher elevations and rocky soils of northeastern Alabama and adjacent Georgia. (Courtesy Anne Runyon)

To live in the Piney Woods, early settlers learned to fence their yards to prevent hogs from rooting in their gardens. They also swept their yards free of pine needles to prevent the frequent fires from destroying their homes. (Courtesy North Carolina Division of Archives and History)

Tar making was one of the earliest industries in the Southeast. Tar burners piled pieces of resin-saturated pine wood in mounds and sweated the tar out in slow-burning fires. (Courtesy North Carolina Division of Archives and History)

Early turpentining was destructive and wasteful. This worker in Ocilla, Georgia, in 1903 is preparing a longleaf pine for turpentining by chopping a "box" in the base of the tree. The box will catch the resin that runs after other workers hack grooves in the tree. (Courtesy Forest History Society)

Turpentine laborers illustrate some of the work that goes into the industry. In the foreground, the grooves that the "chipper" cuts into the wood every week during the growing season have created a resin-coated "face" that gradually rises up the tree. The "dipper" (right) empties the accumulated gum from the box into a small barrel. The contents of this container were dumped into a larger barrel (left) and hauled to the distillery. In the background, a man scrapes the gum that has stuck to the tree. The "scrape" made less valuable turpentine spirits and rosin.
(Courtesy North Carolina Division of Archives and History)

The yard of the H. B. Culbreth distillery in North Carolina, photographed in 1889, was typical of the woods stills then common in the Southeast. The still was generally located near a river, where barrels of naval stores could be floated to markets along the coast. (Courtesy North Carolina Division of Archives and History)

Barrels of turpentine and rosin await shipping aboard schooners docked at Wilmington, North Carolina. These homely products of the longleaf pine forest had a worldwide market during the sailing ship era. (Courtesy North Carolina Division of Archives and History)

Rafts of longleaf pine timber were floated down creeks and rivers for coastwise
shipments to northern cities and transatlantic shipments to European ports.
(Courtesy Florida State Archives)

A logging train heads for the coast near Pensacola Florida in the early 1900s.
The railroad revolutionized logging activities in the Piney Woods.
(Courtesy Florida State Archives)

Longleaf pine logs are piled in an East Texas mill yard in the early 1900s.
(Courtesy Stephen F. Austin University)

Longleaf pine logs await cutting at a Florida mill in 1920.
(Courtesy Florida State Archives)

This stump orchard in Louisiana was all that was left of the virgin forest. Such sights were typical of the cutover lands left behind by cut-and-run loggers.
(Courtesy Stephen F. Austin University)

Without seed sources, natural regeneration was impossible on this cutover tract in Texas. (Courtesy Forest History Society)

An old-timey quail hunt in a longleaf pine forest in Thomasville, Georgia. (Lawrence S. Earley, photographer)

Prescribed fire in longleaf pine forests keeps the brush in check and exposes the
soil to receive the longleaf pine seed. (Lawrence S. Earley, photographer)

Many former longleaf pine sites in the South were reforested with slash pine and loblolly pine, but national forests today are restoring longleaf by removing these substitute forests. (Courtesy Forest History Society)

On the night of September 21, 1989, Hurricane Hugo destroyed thousands of acres of longleaf pine in the Francis Marion National Forest. The storm decimated one of the healthiest populations of red-cockaded woodpeckers in the South. Forest damage is still visible in this photograph taken in 1995. (Lawrence S. Earley, photographer)

Young longleaf thrive in an opening in a restored Louisiana longleaf pine forest.
(Lawrence S. Earley, photographer)

EIGHT

A Reckless Destruction

The damage done to the forests of Georgia
by the turpentine business is simply incalculable.
—New York Lumber Trade Journal (1887)

It took the coming of the railroad for turpentine mania to spread. Before the railroad, rivers had been the true highways of the day, not the poor roads studded with stumps and roots, mired with mud that overturned wagons and sucked the shoes off travelers. If turpentiners wanted to ship their goods to the coast, they'd head to one of the thousands of threadlike creeks and streams that linked the interior to the sea and to the world. In those days no self-respecting community was very far from a river, and at countless landings up and down these rivers farmers gathered, waiting for a freshet or a tide to take their goods to the coastal trading ports.

Negotiating the muddy currents was a flotilla of river craft of all sizes. Flats or flatboats had flat bottoms with board sides that could be poled by several men upstream against a current or tide. The piragua (sometimes called "pirogue" or "periauger"; the word probably derived from the Spanish *perigua*, meaning dugout) was a bargelike open craft fashioned from two sides of a cypress log, hollowed out and laid side by side, sometimes a few feet apart to give it greater size. Boards laid across the logs formed a deck that could be wide enough to carry barrels of tar and gum, or even horses and cattle.[1]

All these vessels depended on the currents and the seasonal freshets in late winter and early spring that could float their goods to the ports, sometimes scores of miles away. During a freshet, the rivers bustled with energy and life. From the deck of his Cape Fear steamboat in 1853, Frederick Law Olmsted noted this colorful picture of commercial life on the river: "In the evening we passed many boats and rafts, blazing with great fires, made upon a thick bed of clay, and their crews singing at their sweeps." Rafts sometimes linked up with other rafts, making huge bateaux. And when a raft reached the port cities, after much travail and lost goods, newspapers reported its arrival:

Sloan's flat arrived yesterday with 6 casks spirits of turpentine, 118 barrels rosin and 34 barrels tar.

Peterson's raft arrived yesterday with 56 barrels tar.[2]

On tidal rivers such as the Cape Fear, once rafts, flats, and other current-driven boats reached the lower sections, they tied up twice a day and waited for the tide to turn, adding hours to a trip. Moreover, the common summer and autumn droughts meant that rivers and creeks were often too low to float goods to market when turpentining activities were at their peak and when hundreds of stills in the longleaf belt were pumping out thousands of barrels of turpentine spirits and rosin. They would have to be kept ashore until a freshet could float them down.

This unpredictable form of river transportation created havoc with the mechanisms of supply and demand, contributing to the volatile pricing that afflicted the industry. "The water courses continue low, and the arrivals of produce very light," reported the *Wilmington Daily Herald* at the beginning of November 1854. Whoever could get goods to port at this time could take advantage of high prices. But the reality was that suppliers could get their goods to Wilmington, Savannah, and other southern ports mainly during the times of the freshets, from November to April. And with everyone trying to sell their products at the same time, the price was tugged downward.[3]

The invention of the steamboat helped only a little. As early as 1818, two paddle-wheel steamers, the *Henrietta* and the *Prometheus*, were in service on the Cape Fear River and another on the Neuse River in North Carolina. By midcentury, twelve of them were plying the Cape Fear. Yet even shallow-draft steamboats could not float where there was no water. Community boosters upstream, desperately desiring steamboat trade to stimulate commerce, dredged impossibly narrow rivers to lure the steamboats, but the shifting sandy bottoms defeated even the U.S. Army Corps of Engineers, and the navigability of these threadlike waterways for steamboats remained a chimera. Larger rivers like the Cape Fear or the Chattahoochee were hardly better. Steamboats grounded easily in summer, and even the smaller steamboats were used to tow flats during this season rather than carry freight themselves and risk grounding. Sectional rivalries developed over which rivers were navigable longer. Boasting of the virgin forests of Florida, Georgia, and Alabama in 1856, one reporter wrote:

The Chattahoochee, it is true, (too often of late) is often too low for steamboats to run. But compare it with the North Carolina rivers and creeks which have heretofore supplied the world with [turpentine]. There they often have to wait for a winter freshet to get their crop to market on rafts. Whereas the Chattahoochee is never (not even now, Oct. 10th,) so low but that rafts of turpentine might be floated out and down to Apalachicola; and N. Carolina experience shows that this mode is the fastest and cheapest to get it to market; it not costing more than one-half of what steamboats can afford to freight it at.[4]

The railroad revolutionized the turpentine industry by enabling a predictable schedule of deliveries and by facilitating the industry's expansion into forests in the interior. The first railroad in the United States connected Charleston and Hamburg, South Carolina, in 1833. In North Carolina, the Wilmington and Weldon Rail Road was completed in 1840 linking Wilmington, in the southern part of the state, to Weldon, in the northern part, and connecting North Carolina to Virginia's new railroad. "The making of turpentine and tar is the almost sole business of the thinly settled population of the pine lands," Edmund Ruffin observed in 1840 after making the trip from Petersburg to Wilmington. The new railroads that began sprouting up throughout North Carolina and other southeastern states opened up vast new areas of hitherto untapped pines for turpentiners.[5]

For some Tar Heel producers, they couldn't have come at a better time. Much of the pine belt was simply not drained by good river systems, and whatever forests were near navigable waters had already approached saturation by turpentine operations. Concern that the industry would shortly fail for lack of new forests had prompted some North Carolina naval stores operators as early as the 1820s to flee the state for fresh forests southward. Now there were more turpentining opportunities at home. In 1853 the Wilmington and Manchester Rail Road connected Wilmington to South Carolina's rich longleaf pines, and three years later the North Carolina Railroad, linking Goldsboro and Charlotte, opened even more forests.[6]

Assisted by railroads and steamboats, turpentiners moved into lands south of the Cape Fear River, where for many years it was thought that the forest made inferior turpentine. Lands despised and considered nearly worthless at ten cents an acre five years before were being snapped up in blocks many square miles in extent.

But increasingly, turpentiners were casting avid glances elsewhere for new orchards. Though the undisputed leader in naval stores production for almost 150 years, North Carolina was finding that it now held only a tenuous lease on what had become a nomadic industry, especially in the years after the Civil War. Turpentiners were gypsies, seduced by the vision of unmeasured and seemingly inexhaustible virgin forests from Georgia to Texas.

As railroads expanded their operations in other southern states, speculators who had bought blocks of land offered them for sale in North Carolina newspaper ads like this one:

Turpentine Lands
The attention of distillers and others interested in the
manufacture of Turpentine is
called
to the advantages offered by the country along the line of
the Brunswick and Albany Railroad.
This road extends from
Brunswick, on the seacoast, to Albany, in Southwestern
Georgia—a distance of one hundred and
seventy-one miles—through
The Great Pitch Pine Belt
in the State, which it traverses from one side to the other.
Fine, open, well-timbered lands, lying immediately
along the line of the road, can
be bought for about
One Dollar and a Half an Acre,
and the same kind of land, situated about five miles
from the road, can be bought as
low
as
Twenty-Five Cents an Acre.
These lands are generally free from undergrowth,
the timber being tall and
thick,
and, with the exception of the pines having more heart,
they resemble very
closely those of North and South Carolina.
As fine a quality of rosin can be made here as in

the Carolinas—a large
proportion of that manufactured early
in the spring grading W.[7]

Often, the railroads themselves sold or leased these lands. Turpentining stimulated the growth of towns and villages throughout the Southeast. Small communities took root along the rail lines, blossomed, and became shipping centers for the expanding industry. Before 1875 Colquitt County in southern Georgia had only three communities, but fifteen years later there were forty-three, most of them servicing the turpentine and logging industries that had just moved in. In 1890 Mississippi had just 24 distilleries to handle its rapidly expanding turpentine business, but by 1899 it had 145.[8]

The new operations needed experienced labor, and the operators found it, not surprisingly, in North Carolina. Tar Heel turpentining methods had become the standard throughout the South—"according to the procedure in Carolina" was the mantra in the journals that described the turpentine business. "There are many hands in North Carolina who tend 7,500 to 9,000 boxes for their tasks," wrote one South Carolina writer admiringly. A North Carolina origin—"[A] genuine North Carolina dipper"—was certification enough. Turpentiners opening a block of several thousand acres in Florida or even Mississippi went to North Carolina to hire labor, sometimes providing their hands with round-trip tickets so they could visit their families at Christmas. For decades, newspapers in Wilmington and Fayetteville reported the comings and goings of these African American migrants, often with some misgivings: "The train which left on the Seaboard Air Line yesterday afternoon carried two coal loads of negro turpentine hands who are bound for the turpentine regions in Florida. . . . Other hands are soon to follow. Turpentine operators in both Florida and Georgia are now in this part of North Carolina hunting for hands. Labor is already scarce and this will make the scarcity all the greater."[9]

The industry was changing in other ways. A new financial system had been set up after the Civil War that fed off the migratory impulses of turpentiners and fueled it. This was the world of the factor.

Factors were the turpentiners' link to capital and finance. There were cotton factors and naval stores factors; sometimes they were the same individual, often northern merchants out of New York who set themselves up in Wilmington, Savannah, and Mobile. Naval stores factors sold the operator the tools that he needed to run a turpentine tract and still, and even the clothes and groceries with which he and his workers were clothed and fed. Banks considered turpentine operations risky and loans were difficult to obtain, but a factorage house would lend an operator money for leasing or buying new land, building a still, hiring labor or continuing his business during lean times, at an interest of about 8 to 10 percent. Without factors, an operator would have had to finance his own enterprise, which could be very expensive. Whether or not a factor took on a new customer was a business decision that depended on the customer's prospects: how many trees he had, what kind of land they were on, how many boxes they would cut, and the availability of a railroad nearby.

Factors also were commission merchants who sold turpentine and rosin for the highest prices and pocketed a portion of the transaction, usually about 2.5 percent. When a customer sent his turpentine to Savannah, inspectors first checked on its quality and quantity. Like tar making, turpentining had evolved an elaborate system of regulations over the years to guard against the ever-present threat of adulteration and fraud. The spirits were taken to the spirits shed and the rosin to the rosin shed. To pass inspection, coopers had to make barrels out of prescribed materials, identified as to the maker and hooped in a specified manner. They could not leak. Rosin was weighed. The gross weight of each barrel had to be about 500 pounds, although by custom it was sold in terms of the older 280-pound barrels. Rosin samples were taken from each barrel and graded by holding them to the northern light and comparing them to standard rosin samples. Turpentine was inspected by color and volume, rather than by weight. Darker shades were penalized or rejected. Inspectors used gauges to measure the capacity of the barrel and to tell whether the barrel was filled to within a gallon of its full capacity. A full barrel contained about fifty gallons of turpentine on the average.[10]

As exacting as these measures seemed, cheating occurred frequently enough to prompt buyers to complain. In 1887 the Boston Paint and Oil Club and a large number of other dealers in New York, Philadelphia, Chicago, and St. Louis endorsed the sale of turpentine spirits by weight, rather than by gauge. Inspectors made sure that the barrels were filled to within one and three-quarters or two inches of the bungs, but if a canny operator coopered one side of the barrel head thicker than the other and made his barrel staves thicker as well, he could save as much as several gallons of spirits. Other dealers adulterated turpentine with petroleum or mineral spirits.[11]

The inspection results were sent to the factors. After the opening of the Savannah Naval Stores Exchange in 1882, the factors met daily and posted their offerings on a board. Buyers offered sealed bids on each offering and the highest bidder was accepted or rejected, if the factor thought he could get a better price later.

Through the factorage system, debt drove turpentining. Turpentine operators owed the factors, and the workers owed the turpentine operators. Indeed, it was by extending the double-edged sword of credit to his workers that the operator kept them in his thrall. In antebellum times, the operator could count on the cheap labor of slaves for the bulk of his workforce. After the war, the workers had leverage they hadn't had before. They were in demand for the repair of older railroads damaged during the war, and for the construction of new railroad lines and sawmills. All of these projects needed labor, and turpentine hands were constantly being lured away.

Operators tried to keep their workers together as they moved from place to place. When a turpentine company moved into a new location, the workers formed their own community, isolated from other communities in the area, and lived in shack housing provided free by the company. They bought groceries on credit from the company commissary, even worshipped in churches erected by the company. Whenever the operator moved, so would his camp—laborers, still, housing, commissary, church and all.

Well into the twentieth century, it was a perfect way to maintain control over scarce labor. Workers who owned only the clothes on their backs, and sometimes not even them, would run up sizable tabs in the camp commissary and, like millworkers everywhere, work for years trying to pay them off. Examples were common of workers who had tried for two years to work off a debt as small as forty dollars. If an employee left to work for another operator, his debt went with him. Turpentine laborers whose fathers and grandfathers had toiled in turpentine orchards as slaves were slaves of another kind, as Pete Daniel has vividly shown in *The Shadow of Slavery*. The credit system led to abuses that were not even corrected after fair labor laws were passed in the twentieth century. Turpentine workers were held in peonage. Workers who tried to escape were captured, hauled back to the camp, whipped, and sometimes even shot. Summary justice was their fate, for the turpentine operator often suborned law enforcement in the Piney Woods.[12]

In the late nineteenth and early twentieth centuries, turpentining was "outlaw work carried on by outlaws," according to one worker. Some camps had brutal reputations where drunkenness and hopelessness were common and killings were standard Saturday-night behavior.[13]

The irony was that a self-congratulatory rhetoric of cultural betterment had accompanied the turpentiners' migration south. By moving into new sections of the Piney Woods, the industry would be an agent in the elevation of the people "into a sober, moral and intelligent class of citizens." Yet with the growing rough-and-tumble reputations of turpentine camps, the opposite seemed to be true. Turpentiners quickly wore out their welcome. A hundred or so black turpentine workers from North Carolina in a rowdy pine woods camp nearby was not something the local white population could shrug off. Fearful residents wore pistols to town, even to church, to protect themselves from the "lawless" turpentine element. Others complained that turpentine camps were drawing away young black labor from the farms in the area.[14]

In the new century, more and more people were looking in concern and even horror at what was happening to the longleaf pine forest in the wake of the itinerant turpentiner, his gang of blacks, and his shadow, the lumberman. As the townspeople's souls shrank away from the scarified trees that looked like rows of gravestones, as they tightened their gun belts against the danger that turpentiners' represented, they were also beginning to realize that something else was happening over which they had lost control. There's a hint of ambivalence in these words of the editor of the *Tifton Gazette* in 1892: "A few days ago the magnificent forest of pine timber in Colquitt County was unbroken and the axe of the mill and the turpentine man had never entered its borders. Now the rattle of turpentine implements and the swish of sawmills are heard throughout the land turning into available wealth the splendid Georgia pine." As the turpentiners moved on, leaving a forest of heavily boxed longleaf pines, some dead and others dying, where catastrophic fires and the ever-present hurricanes took a heavy toll, it wasn't just the loss of tax revenues that local observers feared. Something more unthinkable was occurring—the destruction of the forest itself.[15]

———

At 10:00 A.M. on September 10, 1902, six hundred members of the Turpentine Operators' Association filled Wolfe's Casino Theater in Jacksonville, Florida, for their first convention, a two-day affair. It was the largest assembly of turpentiners and factors ever gathered. Though Savannah was still the major turpentine-exporting port, Jacksonville was just a year away from opening its own port on the Atlantic. Early in the new century Florida was booming with northern capital. The state had sold 4 million acres to private real estate developers in the eighties, and developer Henry M. Flagler, who had already built a railroad south to Miami, was extending it to Key West and erecting palatial hotels for winter tourists from the North. Florida's turpentine exports were climbing. Already its turpentine establishments led Georgia's in capital invested, and it worked almost as many laborers as Georgia. With the St. Johns River draining a vast area of longleaf pine as it flowed north to Jacksonville, the city was ready to claim its destiny as a turpentine port of renown.

Before the events of the day could begin, however, the mayor of Jacksonville, the Honorable D. U. Fletcher, would speak, followed by W. S. Jennings, governor of Florida. Mayor Fletcher's comments were pleasant, even complimentary as he assured the turpentiners that Jacksonville wanted their business and was proud to be associated with the industry. He acknowledged that his remarks were properly limited to this wholehearted welcome, but he couldn't help adding a few suggestions that he hoped they would find "worthy of consideration":

> You wish to pursue the industry in which you are engaged. The people of the State are interested in your doing that, because of the great returns from your product, but particularly because that result must mean the preservation of the pine forests, not only for naval stores purposes, but for timber as well. You seek to accomplish this by restricting production. . . . If you would supplement it by inaugurating and enforcing a better system of opening the trees and gathering the gum, you would, it seems to me, take care of the forests, and at the same time produce what you want. One can ride on the train through the orchards and see that there is a reckless destruction of the trees going on without a material increase in the production over what would be realized if that were stopped, and this means a rapid and unnecessary depletion of the forests. . . . At the present rate your industry will not last fifteen years.[16]

Some audible grumbling from the audience may have ensued, because what he said next has an extemporaneous feel:

> I hear some of you saying: "What does a lawyer know about our business?" You sometimes speak of jack-leg lawyers. I want to be frank and say that there are some jack-leg turpentine operators. You exhibit no more science in the present system in the way you open your trees and gather your gum and destroy the forests, than does the jack-leg lawyer in the handling of his cases. I call on this association to bring about needed reforms for the sake of the forests and for the sake of your own industry.

The newspapers were silent about the reception Fletcher received. If the turpentiners had hoped that Governor Jennings, who followed him to the podium, would lay the balm of concord on the assembly after the mayor's hard words, they were mistaken. There would be no quarter given to the turpentiners that day.

Jennings recounted a lengthy and standard history of the turpentine industry—how it had arisen centuries before Christ and the New World had been blessed with the greatest turpentine forest in the history of man. The turpentiners had moved first through North Carolina and then into South Carolina, Georgia, and now Florida. Here the turpentiners had found apparently inexhaustible resources "which they proceeded to destroy in the most wasteful and reckless manner." His voice rose with tough questions: "Is not an industry of such magnitude as this worth the effort to preserve it? Is it good business foresight to annihilate the source of all these millions? Is it just to coming generations that we should destroy these splendid forests, without an effort to replace them?"

A charged silence must have dominated the hall at these words. But the governor wasn't quite finished:

> In looking over the State as I saw it first, I recall mile upon mile of lofty pine stretching away on all sides, standing like lofty brown columns, supporting arches of living green, through which the breezes, as they passed, made sweet music; but now as I traverse the same country, I see great areas of scrubby pines, jack oaks, and a little wiregrass. The cathedral arches have gone, nature's organ is silent, and the colonnades, with their everchanging vistas are no more. The hand of desecration rests heavily on the bosom of the earth, blackened stumps alone pathetically tell of the monarchs that once made the land beautiful and valuable, and I am told that the ancient order of turpentine men wrought all this desolation. . . . In the track of the naval stores and lumber men there are only blackened stumps to pitifully tell the story of the past, of a beautiful land left a ruin [by] ruthless, wasteful extravagance.

According to the *Florida Times-Union and Citizen*, Mr. P. I. Southerland of Middleberg, Florida, was introduced after the governor took his seat and made a "brief and pleasant response" on behalf of the Turpentine Operators' Association.

Though bruising, the criticisms of Fletcher and Jennings were almost commonplace in 1902. By then, the reputation of the turpentiner had slunk to the level of the Tammany Hall politician. The turpentine boosters of the 1850s had sent their troops marching off to Georgia and Alabama as colorful, if rustic, figures flying the flags of high purpose and noble endeavor. It was industry and prosperity, civilization on the march. The harsher reality was that the itinerant turpentine industry was hacking its way through the Piney Woods "like an army of locusts leaving a Kansas wheat farm," as one critic put it. It was an army that recognized no loyalties of place or region, nor was it interested in sustaining the forest that gave the industry its meager livings. According to the evidence, turpentiners and lumbermen had brought nothing to the forest but greed, left nothing behind but waste and devastation, and pocketed nothing but pennies.[17]

Why was turpentining so destructive and wasteful? Primitive methods such as boxing certainly had a lot to do with it. Boxing did not necessarily kill the tree. It is common to find old boxed and chipped trees that are still living. If a turpentined tree is left healthy by the turpentiner, its bark begins to roll over the bare face like cooling lava. Given enough time, the turpentine face may even disappear entirely. In Florida, the chipped faces of trees worked by the Spanish before 1750 have been found buried deeply inside the tree, with no trace on the exterior of the wounding it once received.[18]

Yet cutting into a living tree with an ax or multiple times with a hack was not conducive to its health. In 1923 a naval stores specialist with the Department of Agriculture, in an article titled "The Goose and the Golden Egg," showed photographs of a longleaf pine stand in which turpentining had been directly responsible for killing 75 percent of the timber. To achieve that mortality rate, the turpentiners must have been spectacularly careless or incompetent, but the spectrum between competence and incompetence in turpentine work was not all that wide, and a lot of trees fell victim to the small difference.[19]

The early booster literature is full of instructions about the correct way turpentining was to be done. A worker should cut a box on the straight side of the tree, no higher than eighteen inches from the ground so it would not be filled with rain, and on the north or shady side of the tree. The box should be only about four or four and one-half inches deep and eight inches wide so as to enter the heart of the tree as little as possible. It had to be cornered precisely to make channels that enabled the gum to run into the box.[20]

Similarly, chippers mustn't chip too deeply. The gum would run just as well with a narrow chip as a wide one, with a shallow chip as a deep one.

And dippers should prevent the gum from spilling. Operators should also refrain from working multiple boxes simultaneously on the same tree.

Despite these injunctions, conservative turpentining practices were uncommon. Frederick Law Olmsted, on his travels through North Carolina in 1853, estimated that "in ninety-nine 'orchards' out of a hundred, you will see that the chip has always been much broader and deeper than, with the slightest care to restrict it, it needed to have been." Slaves had few incentives to do a good job, especially given their workloads. A box cutter was responsible for cutting fifty to seventy-five boxes a day, and a chipper had to work his crop of 10,000 or 12,000 trees every week. Thus boxes were often cut too deep, and the chip was often much wider and deeper than was recommended. Dippers leaked gum all over the forest floor, the equivalent of spilling a can of gasoline in an old barn and hoping it didn't catch fire. Greedy operators might cut multiple boxes in a tree and work them simultaneously rather than successively.[21]

All of these actions could kill trees directly or indirectly. Carelessly deep gashes in the tree could kill trees outright. Boxing made the tree susceptible to insect attack, and it weakened trees and made them vulnerable to hurricanes. During the devastating hurricanes of August and October 1893, turpentiners in North Carolina reported that the unboxed timber had sustained little damage, but that much of their boxed timber had been devastated. "Out of one hundred prostrated long-leaf pines seen near Bladenboro," reported state forester W. W. Ashe, "85 were broken off at the box, four above the box but along the face, 10 were blown up by the roots, and only one tree was broken off above the face." Trees with multiple boxes were even more vulnerable.[22]

Boxes also contributed to the destruction of the tree by fire. A single blaze might not kill a tree, and unboxed trees had impressive resistances to flames, but over time the frequent fires that haunted these forests licked ever deeper into the open boxes and could eventually kill the tree or open it up to a fatal insect infestation. One who witnessed a fire in a turpentine forest was deeply impressed with its destructive power:

> I have seen the mountains on fire—have read of prairies on fire—but a fire in an old turpentine orchard, overgrown with rank grass and weeds—every tree coated with [resin] for ten or twenty feet from the ground presents an awful and sublime sight. I never knew what a power uncontrolled fire was before. The roar of the flames could be heard a great distance—as the fire leaped from tree to tree—20 feet

or more above the ground—like living bodies all aflame. Never read of anything more grand or terrific.[23]

Everyone knew that the prescription for a long-lived turpentine orchard included raking the needles away from the resin-coated trees. But if a turpentiner was a little short of labor, or if he had a choice between raking and cutting new boxes, this job was often left undone.

Boxing was also incredibly wasteful. The whole point of the box was to capture the valuable resin that flowed from the tree after it was wounded, but at that it was extremely inefficient. As gravity carried the sticky gum down the face toward the box, a certain amount stuck to the tree. How much of it arrived in the box as a soft dip depended on a lot of things, but most clearly it depended on how far the wound was from the box. During the first year of operation, when the hacker's chip was only a foot or so above the box, much of the gum flowed to the box. The yield from a first-year tree was known as "virgin dip" and made the most spirits of turpentine and the clearest and most valuable rosin. In subsequent years, as the face, or the area that the chipper worked, rose higher and higher on the tree, less of the gum arrived at the box and what did was darker, because of evaporation and oxidation, and worth less than virgin dip. More and more of the gum stuck to the tree, and at the end of the season this was removed and sold as "scrape." Scrape was worth much less than virgin dip. On a face that was eight or ten feet high, almost all the gum would stick to the tree. At that point, according to the literature, the tree had "run to scrape."

Only virgin forests and trees that had never been turpentined yielded the virgin dip that made the most money. Thus, the longer an operator worked the same piece of forest, the less he made. Given this evidence, the enterprising turpentiner believed that the gateway to profitability lay in abandoning a turpentined forest after one or two years and purchasing or leasing virgin forests elsewhere.

The economic reasoning behind the nomadic existence of industrial turpentining is made graphically clear in *My Southern Friends*, a fictionalized work written by Edmund Kirke (pseudonym for James Roberts Gilmore) that was published in 1865. The narrator is a New York factor who is visiting some of his southern clients. One of them, a man who owns a plantation on the Trent River near New Bern, North Carolina, is in debt to the factor. He is losing money on his turpentine acreage, and although his cotton and corn are profitable, he doesn't have enough land in these crops to offset his losses. His turpentine operation is failing for two reasons, he tells the factor. The trees have run to scrape, and they have been thinned for lumber with the result that his hands spend more time walking from tree to tree and harvest less and less gum.

The factor suggests that the plantation owner can make good money by cutting shingles from his swamp cypress, but the soft-hearted landowner doesn't want to send his slaves into the swamps because the work is too hard. Then there are only two alternatives, says the factor: sell your plantation and slaves, or move to a new location with untapped trees. "Buy 10,000 acres on the line of the Manchester railroad for 75 cents an acre. This railroad is finished to Whiteville [North Carolina] and land can be gotten within 20 miles of Whiteville." The producer accepts these terms and, in due time, according to Kirke, manages not only to pay his debt, but also to run a profitable operation. What Kirke doesn't say is that by accepting this advice, the operator is jumping on a moving treadmill that won't stop until he has either run out of virgin forest or gone bankrupt. He has committed himself and his workers to a nomadic existence with fateful consequences for them and for the forest.[24]

As the century wore on, the pace of forest abandonment picked up. Operators worked and then discarded hundreds of thousands of acres of orchards each year. Why work trees for even four years when virgin dip was the moneymaker and the forests were inexhaustible? It was a general pattern, according to the Tenth Census of the United States in 1880, that "[turpentine] manufacturers, unable to make a profit except from virgin trees, abandon their orchards after one or two years working and seek new fields of operation; the ratio of virgin forest to the total area worked over in the production of naval stores is therefore constantly increasing."[25]

In the deserted turpentine orchards, with the trees weakened by the boxing and hacking and with faces painted with flammable resin, the dual threat of fire and windthrow increased enormously. It didn't help that for many years timber cutters rejected a forest that had already been turpentined because of the prejudice that lumber from bled trees was inferior to that from unbled trees. Through the 1890s, according to one estimate, the loss of timber to fire, windthrow, and disease in abandoned turpentine orchards in the Carolinas, Georgia, Florida, Alabama, and Mississippi amounted to a staggering 18 to 60 billion board feet.[26]

Many have suggested that turpentiners moved on because they were "exhausting" the trees, bleeding them dry in their rush to exploit them. But that doesn't seem to be true in the main. Theoretically, at least, most trees could have been worked for many years, and at one time in North Carolina that may well have been the norm. In 1894 W. W. Ashe, a forest surveyor for the North Carolina Geological Survey, found many orchards that had been worked for decades — from twenty-five to thirty-five years in the Cape Fear region, for example, and elsewhere for as long as fifty years, "with intermissions of a few years for rest and to allow the space between the hacked faces to increase in breadth." But after the industrial model of turpentining took hold, there were few if any references to high faces in the Deep South. In the rush to maximize profits, virgin dip became the lure.[27]

What profits there were, that is. Turpentiners debated the profitability of their industry throughout the nineteenth century. Even in the booster literature of the 1840s and 1850s, dissenting voices could be heard just below the assurances of high profits. Many turpentiners failed to make a reasonable profit, even with slave labor, and believed that the industry was a risky enterprise at best. Before the Civil War, the cost of labor could be ignored, but after the war labor became a significant cost. In 1884 a turpentiner taking a four-year lease on 4,000 acres would be investing $50,000 in land, labor (chopping boxes, inspecting, cornering, raking, chipping, dipping, hauling, distilling, coopering, and superintendence), materials, interest on capital invested, depreciation, taxes, and incidentals. It would cost him fifty-five cents per tree to turpentine the land, and he would produce seventy-five cents worth of products per tree. This left twenty cents net profit over four years, or a nickel per tree annually, a dollar and a quarter an acre, $5,000 for the entire twenty crops.[28]

In reality, however, all turpentine orchards were not statistically identical. The profitability of an orchard depended on many things besides the cost of leasing land and paying and feeding labor. A canny turpentiner would check to see how many boxable trees grew on each acre. In a thin stand, workers spent more time walking and less time in productive work. This increasingly became a factor in the new century as the best plots were worked out or as second-growth forests began to be turpentined. A telling statistic at the end of the century was the change in the definition of a crop, or the number of trees a worker might expect to attend to. In the 1850s, with slave labor and a virgin forest, a crop was defined as 10,000 or even 12,000 trees; by 1935, with paid labor and second-growth forests, the average working crop was estimated at 6,000 to 8,000 trees. Turpentiners also had to be aware of transportation costs. Hauling gum or finished products a great distance overland decreased profits.[29]

The kind of land you turpentined influenced your balance sheet. On some pieces of pineland, the trees just never produced as much as on other land. Even on the same land, neighboring trees produced differently. "It is not yet possible to explain . . . why a misshapen runt sometimes yields more gum than a fine, symmetrically developed neighbor of larger size," one naval stores researcher wrote in 1935.[30]

What was an even more serious limit on profits was a persistent tendency of the industry toward overproduction. So many turpentiners were pushing into ever more virgin forests that vast supplies of turpentine and rosin periodically inundated the market in quantities that couldn't be absorbed. Prices fell. Depressions were announced. At times the price was so low that it wasn't profitable to ship turpentine and rosin to market.

Periodically, turpentiner associations attempted to tame overproduction in a variety of ways, including calling for reductions in the number of new boxes cut. That sometimes had the opposite effect, suggesting to some sharp operators that they could take advantage of the reduced harvest. "The trouble is, however, that about nine out of ten of his neighbors are figuring on the same thing themselves, and over-production is the natural result every year," glumly assessed the Naval Stores Manufacturers' Protection Association of Georgia in 1888.[31]

In 1921 Thomas Gamble, editor of the *Naval Stores Review* in Savannah and an enthusiastic turpentine bull, nonetheless could only mourn the passing of the virgin forests and the slim recompense the turpentiners had received for their work. "One cannot but breathe a sigh of regret for the millions of acres of noble trees that have disappeared, bringing back many claim but a minor stream of gold in replacement, for the industry passed through many years when the returns were in nowise commensurate with the efforts put forth and the great raw wealth that was swept away."[32]

———

Given boxing's reputation for destructiveness and inefficiency, it was scarcely surprising that throughout the nineteenth century many attempts were made to find alternatives. If turpentiners could collect more dip, produce less scrape, and distill better quality turpentine and rosin, they would not have to migrate every two years and they would pocket more profits. As early as the 1850s, one turpentiner nailed a movable receptacle, "a tin box," near the chipping spot to gather the soft dip and prevent scrape. Nothing came of it, but this idea must have been slowly percolating through the minds of many turpentiners of the day. By 1900, among the 102 patents devoted to the turpentine trade, more than a dozen were based on this revolutionary idea. Commercially, they were all destined to be failures.[33]

All but one, that is.

On the evening of September 10, 1902, at the same event at which Mayor Fletcher and Governor Jennings had tongue-lashed the assembled turpentiners in Jacksonville, Charles Holmes Herty took the podium to present his idea for an "Apparatus for Collecting Crude Turpentine," as his patent application described it the following February. Thin and bespectacled, Herty was a thirty-five-year-old Georgian who had been recently hired by the Bureau of Forestry, soon to be the U.S. Forest Service. A chemist by training, he had spent much of his adult life as a professor of chemistry at the University of Georgia. In 1899, frustrated at the lack of opportunities for applied chemistry in the South at the turn of the century, he had sailed to Europe to spend a year studying in Germany and Switzerland. When he wasn't attending lectures, he was making the rounds of various manufacturing enterprises. In Charlottenberg, near Berlin, he made the acquaintance of a famed industrial chemist named O. N. Witt. Innocently, Herty asked Witt his opinion of the turpentine industry in the American South. "You have no industry," the professor is said to have replied. "You have a butchery. I speak from personal knowledge, for I have been in Florida. You are wasting your natural resources and get nothing like an adequate return from them."[34]

Witt's words resonated in Herty's mind, and after his year abroad, he returned home curious about the turpentine industry in Georgia. He found a sense of desperation among many in the business. Herty decided to see if he could find a way to replace the box method with a less destructive substitute. He discovered that the French had been turpentining for decades using a more conservative system. When the southern United States was blockaded during the Civil War, the French decided to develop their own turpentine industry amid the plentiful maritime pines that they had planted in the region south of Bordeaux, a vast sandy plain known as Les Landes. Here they developed their own methods of turpentining. The French used a clay cup and gutter, very narrow faces, and rest periods between working new faces. Each year the cup was raised to collect the gum. With this method, the French had been able to work a single tree for decades rather than a few years.[35]

Herty decided to adapt this method to American forests and American conditions. The method that he had developed, he told the turpentiners, did not involve cutting a box. He hung an earthen cup similar in shape to a flowerpot beneath each turpentine face, with two galvanized iron gutters just above that channeled the dripping resin into the cup. He had tested his cup and gutter system against the older box system on four crops of trees—virgin trees, second-year trees, third-year trees, and fourth-year trees. Half of each crop was boxed; the other half was cupped. The cupped trees had outproduced the boxed trees by 23 percent, and the resin had made lighter grades of rosin. His experiment had shown increased profits of almost $2,000, more than enough to pay for his cups and gutters and still have a little left over.[36]

According to the *Florida Times-Union and Citizen* the following day, Herty's audience was enthusiastic about his invention. It had a profound effect on the U.S. Forest Service as well, which had been among the most critical of the shoddy and wasteful methods of contemporary turpentining. The men and women of this new government agency enthusiastically supported the idea. In February 1903 the Forest Service published Herty's paper, "A New Method of Turpentine Orcharding," and followed it up with the release of circulars to newspapers in the turpentine belt. One newspaper writer had this comment to make:

> Turpentine operators and timber land owners who have seen the results of the cup-and-gutter system are enthusiastic about it. It means much for the turpentine industry; much also for the owners of southern pine lands, who have seen their new timber land laid waste by the destructive boxing. A method of turpentining that inflicts so little damage on the trees is a most important factor in the problem of preserving southern timberlands, and as such it marks another advance in the progress of forestry.[37]

Herty's new method didn't entirely vanquish the old. Boxing hung on in many places, and waste continued to be a specter that haunted turpentining well into the century. Nevertheless, as the industry swung West, the cupping method enabled turpentining to survive against the growing hostility of lumbermen. In the 1890s giant timber companies that had bought hundreds of thousands of acres in Mississippi and Louisiana increasingly refused to lease their land to private turpentine operators, seeing the harm wrought by boxing in Florida and Georgia. With the advent of the cup-and-gutter system, however, they gradually began to turpentine their holdings themselves using the new methods. In 1908, when rosin values for the first time exceeded turpentine values, some in the industry attributed it to the tremendous improvements in the grades of rosin brought about by the new methods of turpentining.

Turpentine and rosin production in the United States peaked in 1909 and gradually diminished to an insignificant amount by the end of the twentieth century. Although turpentining still lingers in a few places in Florida and Georgia today, it has changed profoundly. The U.S. Forest Service attempted to improve workers' techniques and the quality of the products by creating a naval stores experimental station in 1932 at Olustee, Florida, and the Naval Stores Conservation Program in 1936. But turpentining was fighting a losing battle as black labor increasingly turned away from the rough woods work and migrated to the North. Gradually the old gum naval stores industry, with workers chipping trees and then distilling the gum to produce spirits and rosin, was joined by the less labor-intensive wood naval stores industry, which extracted spirits and rosin from resin-saturated virgin stumps and roots collected from the forests. Other attempts to increase worker productivity and save the tree included the use of sulfuric acid rather than hacks to make the resin run. In 1973 the Forest Service closed the Olustee Station. As Percival Perry put it, "Like the village blacksmith, the trail-driving cowboy, and the one-horse shay, the 'turpentine man' had had his day."[38]

Assault on the Southern Pines

There is a portion of Alabama known as "the long-leaf pine region." It is thickly timbered with this valuable yellow pine lumber, which contributes to the supply of nearly every European city, and provides masts and spars—so tall, strong and straight is this timber—for the sailing vessels of almost every nation on the globe. A railroad has just been finished through this section, and it goes without saying that the resources of the country have scarcely been touched.
—Charles H. Wells, *Baltimore Manufacturers Record* (1884)

George Brennan and I got to the St. Marys River a little before noon under a sky that looked like it was going to unload any time. Rick Bennett and Jack Ring were still on the river, so we waited at the truck, with Brennan talking a blue streak. Brennan was fifty-nine at the time. He was once a commercial truck driver, but his arthritis got so bad he had to have a hip replacement that left him with a pronounced limp and an inability to do much driving except for the occasional run to the river to pick up logs.

A couple of locals stopped by. "What are those boys doing on the river with that diving gear?" asked an old farmer wearing blue coveralls and a straw hat.

"Those boys are gonna drag some river logs to the landing," Brennan replied.

The man snorted. "I've pulled thousands of logs from the swamps with nothing more than a team of oxen. There's nothing better to pull logs with than oxen, better than horses or mules."

The two men talked about pulling logs for a while, then traded a couple of stories about hunting with beagles until Bennett and Ring appeared at the foot of the landing, slowly towing a big log that snorkeled just below the surface. Brennan eyed the log suspiciously.

"Now here they come bringin' a monstrosity," he grunted.

He limped over to the truck and backed the converted boat trailer down the ramp until its tires were all but submerged. While Ring positioned the johnboat to keep the line taut, Bennett jumped into the water in his wet suit and directed the floating log toward the trailer where Brennan was unwinding a steel cable from the winch. When Bennett had tightened the cable around the end of the log, Brennan winched the log up onto the trailer as the electric sound filled the air.

The pine log was glistening black, about the length of a telephone pole and maybe eighteen inches in diameter. It had been felled with an ax, and two sides had been squared off to make it easier to stack in the holds of schooners. Floated down the river with hundreds of others, this log had separated from the raft and sunk. The loss of sinkers, or "deadheads," as they were called, was common from rafts, with estimates running from 25 to 35 percent of the total. With great decay-resistant heartwood centers, the trees could stay incorruptible for hundreds of years at the bottom of the river. During the Great Depression, people made money by pulling virgin longleaf pine and cypress logs from rivers and selling them. Now, Bennett and Ring were trying to get what was left. Brennan would drive the old timeworn warriors to Micanopee, Florida, where George Goodwin of the Goodwin Heart Pine Company cut them into expensive pieces of heart pine flooring, stair treads, and siding.[1]

They weren't the only ones scavenging for virgin timber. As Goodwin put it, "They don't make this stuff anymore, you know." Because trees generally aren't allowed to grow old enough to produce high-grade lumber, commercial salvaging businesses have shown increased interest in recent years in recovering old sunken logs from rivers and lakes and milling them for specialty buyers who can afford the high prices. Wildlife interests usually oppose the practice because of its effects on fish habitat, especially spawning areas. Some states, including North Carolina and Florida, grant permits for log salvaging while others have banned it. Many specialists in the heart pine specialty market get their wood by salvaging timber from old houses, factories, wharves, and mills that were built of heart pine, sometimes drawing fire from preservationists.

Bennett preferred river logs to those salvaged from old buildings. "The stuff that's already been out in the atmosphere for one hundred years has lost some of its life," he said later that day. "You have to deal with nail holes and bolt holes and termite damage and beetles. This stuff is absolutely perfect. If I had my choice between a beautiful board made from a beam and a beautiful board from a river log, I'd take the river log every time."

When the rafts were floating virgin logs down the St. Marys, longleaf was being bought and sold under so many different regional names that it became a source of great confusion. It was known as "yellow pine," although in some places it was also referred to as "southern yellow pine," "southern hard pine," "southern heart pine," or "southern pitch pine." In the lumber market in 1897, lumber sold as "yellow southern pine" might have come from longleaf pine, loblolly pine, slash pine, or even pond pine. The lumber sold in England as "pitch pine" or "Georgia pine" was longleaf pine, as was "Georgia pitch pine," "Georgia yellow pine," "Georgia heart pine," and "Georgia longleaved pine." Buyers would demand "Georgia pine" and reject "Florida pine," "Florida yellow pine," and "Texas longleaved pine," not realizing that they were all the same.[2]

Sophisticated buyers recognized longleaf by the tightness of its grain and the proportion of heartwood to sapwood. Longleaf had more heartwood than any other southern pine. A prized 220-year old specimen cut in southern Georgia in the late nineteenth century and displayed in New York's American Museum of Natural History measured 17¾ inches in diameter inside the bark, with only an inch of sapwood. In the old-growth longleaf pine forests, the trees grew slowly, producing a compact and narrow grain—twenty-five to fifty annual rings might be packed into a single inch. Depending on the soil in which it grew, a tree 16 inches in diameter could be 125 years old or nearly 300. The wood possessed a steel-like strength that made these trees some of the finest structural timbers in the country. "In tensile strength [longleaf] approaches, and may surpass, cast iron," wrote Bernard Fernow, first director of the nation's Division of Forestry. "In crossbreaking strength it rivals the oaks, requiring 10,000 pounds per square inch on the average to break it, while in stiffness it is superior to the oak by from 50 to 100 per cent." "Weight for weight, it is stronger than steel," one longleaf booster claimed.[3]

Yet, it wasn't the slow growth of a longleaf pine in the virgin forest or the number of annual rings per inch that accounted for the great strength of the wood, but the percentage of summerwood in the annual rings. Springwood is the lighter portion of a tree's annual rings, consisting of wider cells formed in the spring when the tree begins to pump water from the soil. Summerwood is the darker part, formed in the latter part of the growing season when the cells are narrower. The more summerwood in an annual ring, the denser and stronger the wood.[4]

Bennett took a chain saw and noisily trimmed the end from the log, admiring its tight grain. "That's a grandpa, all right," Ring said, grinning.

Bennett hopped back into Ring's boat and they sped back the way they had come. Over the next two hours, the two would bring back about a dozen logs. At the butt end of one of them was the unmistakable sight of a catface, the scars of a turpentined tree.

This river is a good one for divers like Bennett and Ring, mostly because it's deep. Bennett had dived to sixty-three feet and never less than twenty feet. The old log pullers couldn't quite get their grapplers down to the deepest holes. The St. Marys is a blackwater river, stained with tannins from decaying vegetation, but it's surprisingly clear, unlike some brownwater rivers, so called because of the thick sediment that clouds them. "In some rivers, it's so muddy you can't see the light," Bennett said.

Bennett made a pretty good living from log pulling, although it meant that from April until mid-December he had to board in a hotel for four or five days a week away from his family. It's not easy work and it can be dangerous. On tidal rivers like this one, a diver has to be aware of where he is. Bennett said he spent three or four hours on the bottom every day, and the river could flow both ways during some of that time. Unless he was cautious, tidal currents could carry a diver a long way away from his boat.

"I've got a built-in compass now so I pretty much know from the amount of time I've been crawling on the bottom whether I'm in the middle of the river or getting near the bank," he said. "I usually come up within a few yards of my boat."

When he was not in the water, Bennett researched rivers carefully, visiting local libraries and finding out as much as he could about the river and its early settlements, sawmills, and old landings. "I've hit all the major sawmill sites," he said. "There were probably eight different places where sawmills were located on the banks and near every one of them a pile of logs sank that no one got." The bottom is either white sand or mud or clay or hard lime rock. If a log was resting on rock, it was easier to remove, but often enough a sixty-foot log was buried in sand with only a fifteen-foot stub showing, and he could spend hours unsuccessfully trying to dislodge it.

Once he and Ring located a log, they attached a line to it with a buoy to mark its location. They had rigged their johnboat with a winch mounted just behind the bow so they could pull logs from their resting places on the bottom to a staging area along the shore. From here they towed the logs one at a time to the landing, the log playing out behind the boat like a torpedo, sometimes disappearing below the surface and then breaching the surface in a giant splash.

At his mill, owner George Goodwin walked with me, pointing out the old circular saw, built in 1923 and now mounted under one of his sheds. He is an advocate of the high-quality lumber that he milled, and he is proud that he could match what came out of the virgin forests. "Know how most rooms were finished in those days?" he asked me. "They ordered the floor boards full length—three-inch-wide, quarter-sawn planks—so they wouldn't have any joints in them." A room might be twenty-four-feet long, with twenty-four-foot-long heart pine floor boards. Goodwin's customers, besides the very wealthy owners of private homes, included museums and other public places that were restoring damaged pine flooring installed a century or more ago.

Goodwin loved working with old-growth longleaf and bald cypress. To him, a mature forest had a spiritual quality that no forest today can equal. "You won't find it among a row of planted pines," he said. "To my way of thinking, a mature old forest is worth something far beyond its cellulose value."

He lost his patience when I questioned him about the lumbering of the southern pines. "I look back and I see greed. I see unconcern for anybody but for the immediate person involved," he said. "I see some of the biggest sons of bitches who ever walked this earth raping the forest a hundred years ago and giving it no thought. How could they have pleaded ignorance? I can't believe they could."

At one time timbermen looked at the forests of longleaf pine in the same way that others looked at the buffalo and passenger pigeon, the great whales, and the white pine forests—as immeasurable and inexhaustible. Indeed, for two centuries, until well after the Civil War, timber cutters hardly made a dent in the great forests.

It wasn't that the excellent quality of longleaf pine lumber was lost on the early settlers of the Piney Woods. French botanist Andre Michaux estimated during his travels through the South in 1802 and 1806 that nearly four-fifths of the houses in the Carolinas, Georgia, and Florida had been built with longleaf pine.[5] The export market was also avid for longleaf lumber, especially large timbers for the foundation supports, uprights, and roof timbers needed to construct mills, warehouses, bridges, trestles, and pilings. In the eighteenth and early nineteenth centuries most exports of longleaf went to the West Indies, with a smaller portion going to the Middle Atlantic and Northeast. The English were greedy consumers of these timbers. Michaux noted that the English imported giant ranging timbers from North Carolina to Liverpool, where they were sold as "pitch pine" or "Georgia pine" at higher prices than any other American pine. The standards were so high that port inspectors rejected heart pine timbers less than 10 inches square. Any quantity could be ordered, at dimensions from 20 to 50 feet long and from 10 inches square to 20 by 24 inches. Early southern sawmills could not provide the finished lumber required for building construction, so the squared dimensional timbers were rafted to ports where they were shipped abroad and remanufactured into planks.[6]

Southern and foreign shipbuilders prized longleaf's strong, rot-resistant wood for keels, beams, side planks, and decking. They caulked the seams, of course, with pitch from longleaf. From the first days of English settlement in the New World, tree cutters roamed the woods, North and South, culling the largest and straightest trees for ship masts and spars (booms, yards, gaffs). A 120-gun British ship of the line required a mainmast 120 feet long and 40 inches in diameter that could weigh up to 18 tons. Few longleaf pines were tall enough or wide enough to make the biggest masts, but longleaf was just right for smaller ships' masts, yardarms, and bowsprits. Early in the nineteenth century European and British governments contracted with Gulf Coast merchants to prepare squared timbers and spar timbers for their navies. Longleaf made good spar timber because it was strong and supple enough to withstand stormy seas without splitting.[7]

Up until the Civil War, cutting timber and getting it to market were slow and onerous processes, marked by few innovations. Timber cutters worked mostly along rivers and streams and floated the logs to ports. Their timber-cutting operations gradually radiated out into the surrounding forests for a few miles, and when they'd exhausted the timber on both sides of the river, they moved upstream. Loggers cut trees with a single-bladed ax, felling perhaps twenty to twenty-five trees on a good day. Oxen drawing two 8-foot timber wheels hauled single logs to the river and sent them to a mill in rafts. At the end of the eighteenth century William Bartram witnessed slaves at work along the Savannah River squaring pine and cypress timber in a logging enterprise that was typical for many decades to follow:

The log or timber landing is a capacious open area, the lofty [longleaf] pines having been felled and cleared away for a considerable distance round about, near an almost perpendicular bluff or steep bank of the river, rising up immediately from the water to the height of sixty or seventy feet. The logs being dragged by timber wheels to this yard, and landed as near the brink of this high bank as possible with safety, and laid down by the side of each other, are rolled off, and precipitated down the bank into the river, where being formed into rafts, they are conducted by slaves down to Savannah, about fifty miles below this place.[8]

A timber raft in the Carolinas and Georgia consisted of perhaps twenty separate units of timber. Each unit was made up of squared dimensional timbers or sawed planks cut to order, and it was linked to others fore and aft by a two-foot-long "tie." Rafters cut holes at either end of the ties, and a wooden peg slipped into this hole and through a matching hole on the first and last plank in each unit gave the assembled raft enough flexibility to negotiate the frequent meanders of a southern Coastal Plain river. The raft might carry about 30,000 board feet of logs on the narrower rivers and somewhat more on the wider rivers.[9]

Some rafts also carried naval stores shipments. The rafts floated downstream with the current, requiring the efforts of two or three men to keep them from running ashore. In the South, rafts were the main means of transporting timber to the coast throughout the eighteenth century and much of the nineteenth. Of the forty or more sawmills on branches of the Cape Fear River in 1762, one of them belonged to a plantation called Hunthill that Englishwoman Janet Schaw visited in the 1770s:

> I have been at a fine plantation called Hunthill belonging to Mr. Rutherfurd. On this he has a vast number of Negroes employed in various works. He makes a great deal of tar and turpentine, but his grand work is a saw-mill, the finest I ever met with. It cuts three thousand lumbers a day, and can double the number, when necessity demands it. The woods round him are immense, and he has a vast piece of water, which by a creek communicates with the river, by which he sends down all the lumber, tar and pitch, as it rises every tide sufficiently high to bear any weight. This is done on what is called rafts. . . . In this manner they will float you down fifty thousand deals at once, and 100 or 200 barrels, and they leave room in the centre for the people to stay on, who have nothing to do but prevent its running on shore, as it is floated down by the tides, and they must lay to, between tide and tide, it having no power to move but by the force of the stream. This appears to me the best contrived thing I have seen, nor do I think any better method could be fallen on; and this is adopted by all the people up the country.[10]

Indeed, timber rafting was a familiar scene on southern rivers well into the twentieth century. One twentieth-century account of a rafting trip on the Conecuh River in Alabama to Pensacola doesn't seem to have changed much since Schaw's day:

> Back when I was a young man one of the best ways to earn some money was to take a raft of logs down the Conecuh to Pensacola. They had a saw mill at River Falls where the timber would be squared off and tied into rafts that sometimes had as many as 120 pieces of timber. The lighter weight timbers were tied on the outside of the raft and the heavier pieces placed in the middle; this would balance the raft so that it would float good and be easier to handle. After the rafts were assembled, they would blow the whistle at the mill and that would signal the young men of the community that the rafts were ready to take down river. . . . If the river was down a little, it'd take five days. If it was real low, I've been where it has taken over a week. Depends on how the river was flowing. I have made it in three days and two nights. Sometimes we'd have a little shelter built on the raft but we'd usually tear it down two or three times before we'd get down river. It was the best money you could make back in those days—$25 a trip!
>
> The rafts-were about 60 ft. × 150 to 200 ft. long and at the back of the raft was an opening where you'd put the oar to guide the raft. The oars were pine poles that had a 16 × 20 foot long blade on the end. The handle was 40 feet long and you'd put that pole in the water and twist it and that'd steer the raft on. It was not hard to handle, the water would carry the raft; all you'd have to do is keep it in the river.[11]

Mill owners situated their early mills next to or even on top of a creek or river. Water flowing over a milldam onto the mill wheel turned shafts to produce a reciprocating or up-and-down saw movement. It was a slow way to cut three thousand board feet of lumber a day, but it was better by far than primitive pit sawing which took the considerable daily efforts of two men to produce a few planks. Mill technology developed slowly from the eighteenth century to the nineteenth. The single sash saw eventually gave way to gang saws, or several saws mounted in parallel. The circular saw, with its continuous rotary movement, and single- and double-cutting band saws increased mill output tremendously.[12]

In the colonial period, North Carolina exported between 3 million and 4 million board feet of lumber a year and was the South's leading lumber producer and exporter. A French traveler passing through New Bern in the early 1760s noted that "there is plenty of saw mills in this Country set up at little Expence. Wherever there is water that they can raise to the hight of 5 feet by means of a Dam or breastworks they Erect a mill, if there is a sufficient quantity of water . . . they are allways going, as the Contry is Cover with timber." The U.S. Census of 1840 estimated about 3,500 sawmills in the longleaf pine range, almost a third of which were located in North Carolina.[13]

The gentle gradients of the Atlantic and Gulf Coastal Plains hampered the output of water-powered mills as well as year-round rafting. From a slope of only eight feet per mile, such as existed throughout the Gulf region, for example, advanced engineering skills were necessary to lift flowing water high enough to increase power noticeably. Along the Atlantic Coast, streams dried up in summer and grew shallower, preventing year-round rafting operations. The Gulf Coast experienced more rain than the Atlantic Coast, enabling some rivers to float logs almost year round, but in general the rivers' slight gradients handicapped timber cutters there as well. Spring freshets often destroyed mills.[14]

A more pressing concern for lumbermen in the Southeast, especially in the years just before and after the Civil War, was the inaccessibility of most of the pine forests. Along the Atlantic, small-scale timber operations had gradually depleted the longleaf that grew within easy reach of rivers and creeks. It was possible to use oxen to haul logs to the river from a logging site a few miles away, but at distances beyond that logging costs rose steeply. The coming of the railroad would change everything in the decades after the Civil War, but until then most of the longleaf forest lay vast and unknowable, tantalizing in its potential but just beyond lumbermen's reach.

Longleaf also had an image problem that handicapped its initial exploitation. Though solid in the South and abroad, longleaf's reputation was weaker elsewhere in the country. In many northern and midwestern markets, carpenters complained that the tough wood dulled saw blades and exhausted the sawyers. Some blamed longleaf for warping and splitting easily. Painters said that longleaf wouldn't take paint readily because of its high resin content. Some of these complaints were valid; the very qualities that contributed to longleaf's strength and durability—its greater resin content, its large percentage of heartwood, its heavy weight—made it difficult to work with.[15]

Timber buyers also believed that turpentining weakened longleaf lumber by drawing out its resin. "[Longleaf] yields fine wood [after it is turpentined], excellent lumber, but not the best—as largely drained of its essential oil in the turpentine extracted it cannot be as valuable for timber purposes as the untapped tree," wrote James Averitt, a North Carolina plantation owner. The criticism meant that countless sound logs of excellent quality were rejected at the logging site or mill and burned because they carried the turpentiner's marks. As turpentiners swarmed in the forests boxing and chipping trees in the decades after the Civil War, the complaint, if true, threatened longleaf's reputation as a timber tree. But the complaint was untrue. In 1893 chemists in the Forestry Division of the Department of Agriculture found that turpentining withdrew resin only from the sapwood, not from the heartwood, leaving its durability and resistance to decay unimpaired. It was a finding that the secretary of agriculture pointed to proudly that year: "We may then consign to the rubbish heap of baseless theory this belief which has caused much annoyance to the Southern lumber trade and considerable loss in money and valuable material."[16]

Its crotchety reputation aside, longleaf's major impediment to wider domestic markets was a formidable rival—the vast forests of white pine (*Pinus strobus*) that grew from New Brunswick, across Maine and other New England states, in a wide arc across the southern part of Canada, and down through New York, Pennsylvania, Michigan, Minnesota, and Wisconsin. From the beginnings of English settlement until after the Civil War, it was white pine that held the eyes of commercial-minded lumbermen. Carpenters loved it because it sawed easily, did not warp, and accepted paint. It floated and could be rafted to mills. It grew to spectacular dimensions—6 to 7 feet in diameter and 250 feet high. A single white pine could produce 1,000 board feet of lumber, or in some cases much more. At the sensational end of the spectrum, a single acre of white pine might yield 100,000 board feet, compared to 30,000 board feet reported in the densest longleaf pine forests of the Gulf Coast. It had once provided masts for the British navy, and it furnished cheap building materials for northern cities and towns, and, increasingly, for the new prairie towns west of the Mississippi. Best of all, its forests were interminable, according to one northern politician, capable of supplying "all the wants of the citizens for all time to come."[17]

Yet even as the congressman boasted, in the last decade before the Civil War, lumbermen had already cut out the big pines from Maine and were bearing down on the uncut forests of white pine in the Lake states. By the mid-nineteenth century they were cutting billions of board feet in a gathering crescendo, and many timbermen realized that white pine's days were numbered. In the years after the war, as the U.S. population swelled and the giant American industrial engine began to hum, the lumber industry turned southward and discovered another vast forest that it would liquidate within half a century.

What Michael Williams has called big timber's "assault on the Southern forest" began as the South lay prostrate in the years following the Civil War. Driven by the anticipated demise of the white pine, trainloads of northern timbermen journeyed south to explore a rumored ocean of yellow pine. What they found was a largely untouched forest the extent of which they could hardly get their minds around. The federal government owned 47.7 million acres in Alabama, Arkansas, Florida, Louisiana, and Mississippi. Before the war, it had offered unrestricted land grants to individuals for as little as 12.5 cents per acre, but few buyers then were interested in the heavily forested cypress and longleaf pinelands. The Southern Homestead Act of 1866 restricted land purchases to eighty acres a person to discourage land speculation, to encourage small farmers, especially former slaves, and to handicap the defeated South—former Confederates were barred homesteading privileges entirely. Some railroad and lumber companies were able to bypass the law by illegally settling individuals on eighty-acre blocks and amassing large acreage in this manner, but in the lean years after 1865 southern politicians were eager to lay down the welcome mat for redevelopment, and they were among those who successfully fought for the act's repeal in 1876. Unrestricted amounts of state and federal land now lay open for purchase. The states passed generous tax incentives to railroad companies, awarding them large land grants to the lines along their rights of way. Texas awarded railroad companies 10,240 acres for every mile of track it lay down. A rapacious band of speculators and lumbermen responded, flooding the woods and buying up forest blocks the size of entire counties. "The woods are full of Michigan men bent on the same errand as myself," wrote one in 1882. Indeed, of the seventy-three individuals purchasing at least 5,000 acres in Mississippi and Louisiana between 1880 and 1888, slightly more than half were from Michigan.[18]

Although some bona fide lumbermen were involved in the initial land purchases, most of these men were speculators. Individuals and syndicates bought huge acreages for $1.25 an acre, betting that the value of the land and the trees would only appreciate in time. "Northern capital is seeking investment more and more in the Pine lands of the South," reported the *New York Lumber Trade Journal* in 1887. "These lands are enhancing rapidly in value, and investors who are able to hold can expect to realize in a few years a handsome profit."[19]

Between 1876 and 1888, when limitations were reimposed on land purchases, private landowners, many of them from outside the region, had grabbed nearly 5.7 million acres of federally owned southern land. The concentration of pineland in individual ownership grew even more pronounced in the years following. By 1914 the Bureau of Corporations found that 925 people owned nearly 47 million acres in the South, which represented nearly 80 percent of the region's privately owned pineland. If southern politicians had thought that opening the region's timber cupboard to the highest bidder would result in an economically resurgent South, they were wrong: most of the riches went to northern capitalists. These men wanted in on the longleaf hoard because they knew that the demand for lumber was growing each year. Between 1880 and 1910 the population of the United States nearly doubled—from 50 million to 96 million. Immigrants and other pioneers were flocking to the western territories and their treeless prairies, and they clamored for wood to build houses and factories. In Texas, oxen were employed to haul lumber hundreds of miles to new towns, but it was hugely expensive. Meanwhile, the vast Piney Woods of East Texas lay largely untouched.[20]

With an overwhelming demand for a scarce resource, railroad companies could finally justify building lines through the trackless forests of southern longleaf pine. As northern timber companies bought colossal tracts of longleaf pineland in Mississippi, Louisiana, and Texas, a spider's web of tracks swiftly crisscrossed the region; 23,000 miles of rail lines were laid in the South between 1880 and 1890. Main lines connected distant points, and narrow-gauge spur lines radiated into the forest on both sides of the main lines. Sawmills were sited every few miles along the tracks and communities sprang up around the mills, leaving once-thriving river towns dying in their wake. Some of the richest longleaf pine forests in the South now lay open to gangs of timber cutters who attacked them in waves of implacable efficiency.[21]

Over the years logging operations had become highly organized, as a variety of specialized jobs evolved to exploit the forest quickly and efficiently. Civil engineers arrived at a logging site first, deciding where to place the spur tracks. Next the grading crews cut trees, blew up stumps, built bridges, and lay the roadbed for the spur line. Then, the steel gang lay the spur tracks off the main line, and finally the logging crew arrived with their double-bit axes and crosscut saws. The double-bit ax had only been introduced in the 1870s; a decade later, the invention of the crosscut saw enabled a logger to more than double his output. When the loggers had cut the trees on both sides of the spur, teamsters, driving oxen or mules, "skidded" the cut logs to the cars where they were loaded, taken to the mill, and dumped in the mill-pond. After the logging crew finished an area, the steel gang pulled up the spur tracks and laid them down again a quarter mile farther down the main line. With this organization and technology, railroad loggers could finally exploit forests that had been unreachable before.[22]

In the 1890s the skidder entered the southern forests. This steam-powered machine, together with the railroad and the steam-powered sawmill, was to have profound effects on both the rate of cut and the fate of the longleaf pine forest. The skidder had been developed in the Lake states but came into its own in the South. It consisted of a steam boiler, a large revolving drum around which four steel cables were wound, and large pulleys over which the cables were run. It was mounted on a railroad car that hauled it to the logging site. There, horses pulled the cables into the forest on both sides of the tracks sometimes a thousand feet away. Loggers attached the tongs at the cable ends to the logs and the skidder rewound the cables over the drum, pulling the logs back to the cars. With a steam skidder, a logging team could clear-cut and load more than five hundred trees a day — 100,000 to 200,000 board feet.[23]

By 1892 the annual longleaf cut was estimated at 7 billion board feet, and the *American Lumberman* was crowing that "the timber material par excellence of the whole north is longleaf yellow pine." The port of Pascagoula, Mississippi, increased its timber and lumber shipments from 60 million board feet in 1880 to 130 million in 1893; Mobile's shipments soared from 22.5 million board feet in 1880 to 150 million board feet in 1896. At the end of the nineteenth century, the longleaf pine forests of the Southeast were the epicenter of the American lumber industry. Millions of logs streamed out of the Piney Woods drawn by river currents and railroad locomotives. From southern mills—some of them the biggest in the world—gushed squared timbers, planks, scantlings, and dressed timber by the millions, destined for eastern port cities, western prairie towns, South America, and Europe. Southern yellow pine had become the biggest lumber export in the country.[24]

The turpentiner was living on borrowed time. The railroads that were even then revealing fresh forests of longleaf to turpentiners were also unveiling the untouched forests to a race of lumbermen the likes of which the region had never seen. To extirpate longleaf from its northern range in Virginia and northeastern North Carolina took two hundred years; it would take only fifty more to cut the rest.[25]

I had to go to Alexandria, Louisiana, before I understood how that happened. Alexandria is in the west-central part of the state, along the eastern edge of the region once known as Calcasieu. Calcasieu was a symbol of the longleaf yellow pine lumber trade. Between the Sabine River, tracing the border of Texas and Louisiana, and the Calcasieu River, and south of the Red River, enormous forests of huge trees—perhaps 4,500 square miles in all—grew densely. These forests produced from 12,000 to 30,000 board feet of lumber per acre to the astonishment of the first Michigan land buyers. In time, Calcasieu became a trademark of all that lumber buyers wanted from longleaf timber—strength, durability, and great size.[26]

In Alexandria, I found a place that connected me to that moment in time when the enormous forests of longleaf pine in the western Gulf states met up with the new technologies of timber harvesting. Within a forty-mile radius of Alexandria at the turn of the century, seventy sawmills cut longleaf pine, some of them day and night. Most of the mills were gone in a decade or two, abandoned or moved by the mill owners and left without regret by the workers and their families who had moved just down the road to a sawmill that hadn't yet reached the end of its cut. Logging was a migratory life similar to turpentining, and it, too, left scars on families that faded in time to gentle nostalgia. Deeper and more indelible scars remained on the landscape from thousands of miles of logging spurs that snaked through the pines, leaving runelike patterns for later geographers to decipher. In truth, there is little left from that era, except the occasional spur track just visible through the roadside foliage or the charred remains of a log car covered in vines. The mills themselves are long gone.[27]

All of them, that is, except one. On a hot April morning I drove south from Alexandria along Highway 165 past the community of Forest Hills, which bills itself as "The Nursery Capital of Louisiana" because of the seventy-some nurseries that crowd the rural road, to the village of Long Leaf. There I found the sign for the Southern Forest Heritage Museum, and I drove in.

The museum is a group of buildings on a 56-acre lot that includes the essential parts of the Crowell Long Leaf Lumber Company established by Caleb T. Crowell in 1892 at the beginning of Louisiana's lumber boom. The 165 houses for the 300 millworkers are gone, but remaining are the rusted and time-blasted remains of a steam-powered sawmill, planer mill, dry kiln, and other mill structures, plus a steam-powered logging locomotive, a steam skidder, log loader, spur tracks, and other relics of the period. Crowell purchased about 20,000 acres of virgin longleaf at the beginning of the timber cutting, but unlike most of the other mills his mill enjoyed a surprising longevity, running almost continuously from 1892 until 1969, when it closed.

Just a mile or two south of Long Leaf was the sawmill town of McNary, and just south of it was the town of Glen Mora, where some old-timers remember two or three mills, and Oakdale which also had several. North of Long Leaf were Woodworth and Pawnee where sawmills once ran. All of them had exhausted their forests earlier in the century and had gone out of business. In the early 1920s the mill owners at McNary moved the entire town of three thousand people and the mill equipment on three trains to northern Arizona, where they built another mill and began to cut the Ponderosa pines. They named their new town McNary, too. The Crowell mill had outlasted them all by decades, and today it's a state historic site and a museum.[28]

I toured the mill complex with historian Stuart Hanks, a thin, soft-spoken man with a thick, drooping mustache, and with director Don Powell. What did the old buildings mean to the people of Louisiana? More to the point, what lesson would ninety schoolchildren take home with them after visiting the mill the next day?

Powell gave me one answer as we walked:

"You know, people don't understand the South. They think of plantations but less than two percent of the people participated in that. I'm doing some genealogy for my family, and everybody—and I mean everybody—was a farmer. They grew cotton but their land got its fertility depleted—there ain't nothin' that eat up land like cotton; it sucked the nutrients out of it. Now the sawmill people come in. Some of them are pretty tough business people, but they were willing to risk their capital and that says a lot. 'I'm willing to put my capital out here, build this mill, and believe that I can hire people and buy enough timber to make it and sell it at a profit.' And by doing so they provided the means by which southerners had their first real opportunity to participate in the Industrial Revolution. There was no industry down here, a little textile in the Carolinas, and there was a splash of it in Mississippi, but not much. You can't really say a cotton gin was the Industrial Revolution because it only operates two months out of the year and it didn't take that many people to do it—ten people could run it easy—whereas it took 250 to run a sawmill and logging operation.

"So here we are in just this area right here, ten mills just a few miles away from each other. That's 2,500 jobs. The people came from all over, from all these worn-out farms. They were leaving a house where they were feeding the chickens through the cracks in the floor almost, coming to a warmer house, a tighter house, a nicer house, in a community."

A community with a lighted tennis court, Hanks added, with a beauty shop for the ladies and a school for the children. Sawmill villages had baseball teams. Their commissaries offered more items for sale than a country store.

The sawmill, in other words, fulfilled southerners' aspirations for a better life. It represented a transition from the agrarian South, where you worked for yourself and observed the seasonal schedules of farm life, to the industrial South, where your day was regulated by the clock and punctuated by ear-splitting whistles — whistles that got the tree cutters off on the log trains at 5:00 A.M., whistles that started the millworkers' day two hours later, whistles at noon, whistles that announced the arrival of a load of logs or a hurt logger, and whistles to bring the workday to a close. The mill represented an introduction to twentieth-century life.

But the sawmill represented something more, something hard and obdurate and almost frightening. I didn't understand that until I had walked through the sawmill with retired millworker Jimmy Rhodes, who helped me imaginatively kick-start the silent machinery back into life. Rhodes volunteered now at the Southern Forest Heritage Museum, taking visitors like myself on tours of the old mill.

The Crowell mill was a generic type that was being built in the hundreds across the South at that time. It produced 75,000 board feet per day, a modest output by the standards of the time. The Crowell mill was essentially a two-story affair. On the ground floor a Corliss steam boiler drove a central drive shaft that powered numerous belts and pulleys that ran up to the second floor where the milling took place. In a small room above the second floor, the saw filers plied their trade. The belts drove everything — the "jack-ladder" that hoisted the logs from the log pond up to the second-story log deck; the chain conveyors that took the rough lumber to the green shed; the rollers that conveyed large timbers to the timber dock, where they were loaded onto railroad cars; the conveyor that disposed of the edgings and waste products.

The moving belts also drove the men. All day long, a rolling forest of logs made their way, one after the other, from the log pond up to the log deck, where the cutoff man lowered a three-foot circular saw to cut the log to a desired length. Then the deck man threw a switch and shunted the log to one of two sides of the deck. A big log longer than twenty-eight feet went to the "long side" of the sawing platform where it was squared off, destined for the lucrative export trade to Europe. "I once saw two timbers, special ordered, fifty-feet long and twenty-four by twenty-four," Rhodes said. "Usually forty-four feet was as long as they cut." Another log might be instantly judged as suitable to make railroad stringers, twenty-eight-foot long timbers used for constructing trestles and bridges. That log was cleared to the "short side" of the platform where the sawyer would flip it to position it correctly in the carriage. Based on the commercial orders posted in chalk on a board in front of him, the sawyer would make instantaneous decisions about the best use of each log—whether it was a ten-inch by ten-inch squared timber for heavy construction work or a two-inch by twelve-inch board for interior finishing. After the block setter had clamped the log down, the sawyer pulled a lever that hurled the carriage at speeds of up to sixty miles an hour against a large circular saw, later replaced by an eight-foot band saw. In successive passes, the log was slabbed off on two sides to square it up and then sawed into boards.

The sawmill was a tumultuous scene of huge machines moving violently at great speed and hurling two-ton logs about like straw, where saws shrieked piercingly amid the rumble of moving belts, chains, and pulleys. Men were deafened by the din; sawyers and their assistants mostly resorted to hand signals to communicate. One observer described the mayhem in this way:

> The mill resounded with the roar of machinery, punctuated by the special sounds of log-carriages shooting forward and back and the great saws whirring through the logs. These saws bit through a two- or three-foot log in a matter of seconds, or sliced it lengthwise in much less time than it takes to tell it. They cut with deafening, high-pitched screams, and dazzled one with their swiftness. The machinery was so strong in its attack upon the logs that the whole mill actually shook.[29]

Edger saws trimmed the boards of their irregular margins and cut them into standard widths. The slabs and edgings dropped onto a chain-driven belt that brought them to the waste heap to be burned or dropped onto another conveyor on the first floor that carried them to the boiler. The rough-cut boards moved to the "green chain," a series of conveyors that carried them crosswise to the trimmer, which removed knotty sections and cut the boards into standard lengths. The boards were now ready to be sent to the dry kiln to reduce the moisture in the green wood, making the lumber easier for carpenters to work and cheaper to ship.

As Jimmy Rhodes patiently explained how a log rose dripping from the log pond and emerged twelve minutes later as pieces of rough lumber or squared-off construction timbers, I began to understand the remorseless hunger of the steam-powered sawmill and what it meant to the longleaf pine forests of the Southeast. The sawmill ate trees and consumed forests. To justify the huge expense incurred by their railroad tracks, locomotives, cars, sawmills, planing mills, dry kilns, and other costly investments, saw-mills needed phenomenal amounts of land—at least twenty thousand acres of untouched forest to achieve a ten-year run.[30]

Probably the biggest mill of all was the Great Southern Lumber Company, built in 1907 in Bogalusa, Louisiana. Its owners accumulated 600,000 acres of timberland in western Mississippi and eastern Louisiana. With a work crew of 2,500, it turned out 600,000 board feet per day in 1913, and on one record-setting 22-hour day in 1915 it produced a staggering 1 million board feet. It once produced 175 million board feet in a single year, and in its 30-year history it produced 5 billion board feet of lumber, "an amount of lumber that would have built one-half million four-room houses or a four-foot sidewalk long enough to encircle the globe five times," according to Nollie Hickman. It was so huge that a town of 15,000 grew up around it, and it cleared 25,000 acres a year of mostly longleaf pine to keep its mills going. No wonder that by 1900 Louisiana was cutting 3.5 billion board feet a year, most of it longleaf pine. By 1920, 352 southern mills could produce more than 10 million board feet per year. From 1905 to 1925 Louisiana loggers clear-cut 4.3 million acres of virgin timber.[31]

All those belts, chains, and pulleys clanking inexorably hour after hour, year after year, represented an industrial force that had never been seen in any forest before, an industrial might that was more than a match for the enormous forests of longleaf pine that they were pitted against. Only this kind of technology could have chewed up this forest and spat it out again in the form of millions of miles of close-grained lumber and the most impressive large-dimensional building timbers the world has ever known. Only this kind of technology could have cut to pieces by 1925 a forest that had once stretched over the entire Southeast, leaving a desolate landscape, abandoned mill towns, and a dazed population that was just beginning to understand what had happened.

———

It was one thing to see the forest disappear, quite another for a growing number of critics to realize how much of it had been wasted. Lumbermen in the southern forests practiced a form of logging that has been condemned as "cut-and-run" logging; even today it is considered the most shameful chapter in the history of the American timber industry. Aided by the voracious steam skidders and colossal mills, railroad lumbermen cut everything in their path as they powered into the remaining blocks of virgin pine in Mississippi, Louisiana, and Texas. One forester walked a 100-foot-wide forested transect over four miles in southwestern Louisiana and found only a single living pine four inches in diameter on every five-acre block. Most of these pitiful trees were so small that they were incapable of producing seed. The skidders may have been efficient at harvesting a forest, but they killed tremendous numbers of smaller trees, saplings, and seedlings as the line snaked the logs back to the tracks. A logging operation also left behind a flammable tinder of logging debris, abundant fuel for fires that burned with unnatural ferocity. About 30 million acres of cutover pine forests were left in the condition that William Faulkner memorably described in *Light in August* (1932): "a stumppocked scene of profound and peaceful desolation, unplowed, untilled, gutting slowly into red and choked ravines beneath the long quiet rains of autumn and the galloping fury of vernal equinoxes." [32]

Lumbermen cut and ran to avoid losing a forest to destructive fires fueled by turpentine waste and to avoid paying onerous taxes on standing timber. Local governments taxed standing timber at a higher rate than cutover land. Tax men considered even a few small trees standing timber, so mill owners clear-cut the forest, burned it, and then either sold or abandoned it. Severance taxes — a more progressive tax paid when the timber was cut — were not instituted until a decade or more after the last virgin forest disappeared. [33]

Enormous waste in the cutting, milling, and even use of the trees accompanied the rapid destruction of the southern pine forests. Loggers often cut high stumps two feet or higher off the ground instead of the recommended eighteen inches. On a property in South Carolina with about 40,000 acres of timber, one forester calculated that the difference between stumps two feet high and one and a half feet high was nearly 5.5 million board feet of lumber. Careless loggers often reduced a log's value by not leaving the standard three- to four-inch trim length, or leaving knotty or rotten sections in the middle of a log rather than at the end, so they could be trimmed off, or leaving too much merchantable timber in the tops (which were normally left in the woods after the logging operation). On a nineteen-and-a-half-acre tract, one critic complained, close to 2 million board feet of merchantable timber had been left in the tops, at a loss of almost $15,000. Nearly a year of cutting could be added to a twenty-year lumbering operation if all the merchantable timber had been used correctly. [34]

Mills wasted vast amounts of lumber in inspectors' belief that the forests were inexhaustible. Standards were so high that only the heartwood logs with no more than an inch of sapwood and with no knot or blemish fetched the premium price; everything else, including younger and sappier trees that were more than a century old, sold at a fraction of the top price or were burned on the slab heaps. One logger remembered that he could get a good straight log without knots that would measure forty to fifty feet long. "Above this, the log would begin to have knots caused by limbs, and one couldn't give away a log with knots on it. My father said he would set fire to this remaining part of the tree to get it out of the way after the main log had been hauled out. He said he burned up a fortune in timber, but there just wasn't any market for it back then."[35]

As Leonard Slade, a former lumberman from Purvis, Mississippi, told me: "What they throwed away back then is better than what we have now."[36]

Technological advances sometimes brought increased waste. The wobbly blade of a circular saw cut such a wide kerf that it rendered as much as a third of every thousand board feet into sawdust. Competitors joked that circular saws were "sawdust-making machines" that produced boards as by-products. Massive amounts of longleaf pine heartwood were wasted on uses for which a poorer timber would have been better substituted. To construct a single mile of railroad line, the builders used 3,000 heart pine railroad ties seven and a half feet long by nine inches broad and seven inches thick, which might last only five or six years on the damper soils. "Hence, for the construction of the 3,240 miles of railroad traversing the forest of longleaf pine east of the Mississippi River," forester Charles Mohr wrote, "nearly 10,000,000 ties have been required" from several million of the best trees. In 1907 railroad companies nationwide purchased more than 34 million yellow pine ties. In lumbering operations, steel gangs put down and took up the ties so many times that they were often destroyed quickly. Trainmen sometimes burned heart pine logs as fuel in train fire boxes.[37]

Lumbermen often cut more lumber than the market could absorb, depressing prices. Waste and overproduction beset lumbering as well as turpentining at the close of the nineteenth century, contributing to chaos in the marketplace and reducing profits. In the early eighties the editor of the *Daily Register* of Mobile, Alabama, chided southern lumbermen for foolishly selling lumber too cheaply. "What is the cause then of yellow pine not commanding the price that it should command in these markets?" he asked. "It may be found in the fact that it is pushed on the market to dealers and consumers at any price they are willing to pay for it." If lumbermen couldn't make a good profit on their lumber, "they should shut down the mills, hold their lumber and save their timber, and wait for the reaction sure to come."[38]

Lumbermen were not blind to these problems and sought to tame them through the creation of early trade groups such as the Southern Lumber Manufacturers' Association (later the Southern Pine Association). Their goals were to corral their fractious and independent-minded members and to control production and thus prices, but they failed in the main. Their price-fixing efforts ran afoul of antitrust legislation, and there were large-scale investigations of the lumber industry in 1906 and 1907. Lumber piled up in lumberyards throughout the South. Colossal amounts of forest were wasted in this way, cut without a sufficient market for the product and sold at ruinous prices.[39]

Lumber industry representatives looked on the approaching end of the longleaf pine forest with a kind of fatalism. They admitted that millions of acres of longleaf had disappeared with little to show for it, but they believed it was inevitable. One summed it up this way: "The advancing wave of humanity demanded homes and . . . while the lands were being cleared thousands and tens of thousands derived a livelihood from turpentining and saw milling." Some were remarkably unconcerned by the criticism. If they had been asked, they'd say that they had more important things to worry about, such as payrolls and taxes. Addressing the 1893 American Forestry Association congress, J. E. Defebaugh, the powerful editor of the *American Lumberman*, candidly defended the lumbermen's lack of concern. "To those who attend forestry congresses or read the literature concerning the subject no phrase is more familiar than this, 'the ruthless destruction of our forests,'" he said. "That word 'ruthless' seems to define the mental and moral attitude of the lumberman towards the subject under consideration. So violent and unreasonable, in many cases, have been the charges against him that the lumberman has often been forced to an attitude of apparent hostility, that misrepresents his real feeling, which is one of entire indifference." The lumberman was only a cog in the commercial and industrial system, he went on to say. If you wanted to blame someone, blame the voters who supported and maintained this capitalistic system.[40]

At the same congress, the editor of the *North Western Lumberman* shrugged off the devastation that even then was being registered in the South. Lumbermen are single-minded businessmen who are in it for profit, he said. "The saws are run strictly on business principles and for the purpose of putting the last dollar possible into the pockets of the men who operate them." Lumbermen were only concerned with the present, not the future.

Virgin forests were like drugs that addled the minds of lumbermen; good forestry would have to wait until the original forests were gone.[41]

―――――――――――

By 1925 the nomadic lumber industry was in the process of making yet another shift. Mill owners in Mississippi and Louisiana were abandoning sawmill towns with a speed that dismayed onlookers, and buyers were looking for new timber stocks in the Douglas fir forests of the Pacific Northwest.

Loggers and their families were often appalled at the destruction they had wrought. They remembered the towering trees beneath which they had lived for a while. "There were trees right up to where we lived," one resident of Trout, Louisiana, remembered. "They were the biggest, prettiest pine trees you ever saw. Same way around the school. All in front, around the sawmill and on up the hill. They finally cut around the school. I can remember that feeling how everything opened up and it was suddenly so bright. It hurt my eyes." For those who witnessed the contrast, nostalgia was the emotion of the day. "In 1864 when I first went over the railroad from Savannah to Thomasville there was an almost unbroken forest of magnificent pines extending from Bryan to Thomas County . . . but now one may go over the same rout [sic] and scarcely see a merchantable pine," remembered one traveler in 1901. "Sometimes, when I ride along the new three lane highway, or walk on the paved street, I can but think of the many times I have walked over this very hill where the high school now stands," wrote one old woman in 1949. "Instead of paved streets, I walked on a soft path of deep and fragrant pine needles and I needed no umbrella to protect me from the sun, for the dense shade of the towering pines made it almost like dusk, even at mid-day."[42]

For a new generation that had never seen it, the virgin forest became a powerful symbol of environmental degradation. North Carolinian B. W. Wells summed up his sense of the loss in 1932: "Like a fine old aristocracy destroyed by war, the original longleaf pine forest of our sandhills has been completely cut and burned until the entire scene has so changed that no person of the rising generation may now gain any real idea of the majesty and glory of the original forest."[43]

PART THREE

Forest Management

TEN

Forestry Practice and Malpractice

The plain truth of the matter is that in county after county, in state after
state of the south, the piney woods are not passing, but have passed.
—R. D. Forbes, "The Passing of the Piney Woods" (1923)

The continuance of unsatisfactory regeneration in longleaf forests may
be attributed largely to lack of management or to unwise management.
Indeed, mismanagement has been the rule rather than the exception,
due to ignorance of the unique life history of the species.
—W. G. Wahlenberg, *Longleaf Pine* (1946)

As the loggers and turpentiners gorged at the diminishing forests of longleaf pine, professional foresters stepped onto the southern stage. Many of them, if not Europeans themselves, had been trained abroad and were full of European ideas about forests. Forests were finite, they said, but they were also renewable, and a renewable forest could produce a sustained yield of forest products forever. The lumbermen then rampaging through the southern pineries were operating on a different set of assumptions—that forests were inexhaustible and when you cut one down, you merely moved to another. The belief that a permanent forest could provide a sustainable flow of wood was revolutionary, spawning the science of forest management and a cadre of confident forest managers.

Two bedrock principles guided forest managers of that day. The first was that trees were a crop, a commodity to be grown, harvested, and regenerated like any other useful and necessary food crop. It was an enlightened metaphor compared to another comparison of the day, that the forest was a mine from which loggers extracted the trees and left the land, which was the most valuable resource, to be converted to agriculture or another use. If a forest could be viewed as a mine, it was ephemeral. Once its trees were removed, the loggers—mere labor—would move on to the next forest. If the forest was a crop, on the other hand, it could be permanently managed and the work of loggers, now skilled technicians directed by wise forest managers, would ensure that a new crop came in behind the old. Cycles of forests would follow each other, guided by the benevolent hand of man.

"The idea of a crop involves sowing and planting, cultivation and harvest," B. E. Fernow, the first director of the U.S. Department of Agriculture's Bureau of Forestry, explained in 1893, "and when the crop is reaped we expect that it be reproduced and the soil bring forth another crop, as good or better than the first."[1]

Better than the first? Indeed, most Americans of the day did not view old-growth forests with the same reverence as we do today. They were always a bit ambivalent about them, regarding them mainly as impediments—to the construction of a house, to a cotton crop, to easy travel. Foresters were leery of them for another reason: while the virgin forests existed, their excellent ideas would receive little attention. Virgin forests made lumbermen disregard the discipline of scientific forestry. From the beginning, foresters considered virgin forests simultaneously beyond their scope—"Only God can make a tree," as Joyce Kilmer famously wrote—and beneath their talents. "The forest primeval contains much material which is of little or no value," Fernow wrote in 1890. "When at last the stores of the virgin forest are exhausted and it becomes necessary to apply human ingenuity and management to the production of desirable quantities and qualities of wood material on a confined area, a new industry, forestry or forest management, arises." Nature made virgin forests; humans would make the second forest and would improve on the first. Early foresters pinned their hopes for an improving forestry on second-growth forests, not on the virgin forests.[2]

Forestry's second bedrock principle governed the harvest of the crop: it should ensure, not impair, the ability of the next forest to regenerate. Take the trees, in other words, but leave the forest. This, of course, was what the cut-and-run loggers were not doing. Their method—clear-cut, burn, and abandon—was akin to a farmer eating his seed corn.

The foresters and their new ideas of forest management were a lively presence in the American forestry journals of the day, but their influence at first was slight, their voices drowned out in the general din of exploitation. Until the 1890s, a professional forest service on either federal or state levels had not existed. Forestry advocates—like Germans B. E. Fernow, Charles Mohr, Carl Schurz, and Carl Schenck, and Americans Gifford Pinchot and Henry Graves—were basically cheerleaders for the new science of forestry. Until the early years of the new century, foresters languished in an obscure division of the Department of Agriculture known as the Forestry Bureau. This agency served as little more than a "bureau of information," as Fernow put it. Among its early milestones was the massive survey of the southern forests, *The Timber Pines of the Southern United States*, compiled and written mostly by Mohr and published in 1897.[3]

The federal foresters were essentially powerless, and in their impotence, they saw themselves at first as street-corner Jeremiahs crying, "Repent! Repent! The End Is Near!" In North Carolina, forester W. W. Ashe was full of hopelessness. It is likely, he wrote in 1894, "that this valuable tree, the longleaf pine, will become extinct in North Carolina unless some steps are taken to secure its more general propagation." Only decades separated the pines of the South from utter annihilation, he believed. Each report of the Bureau of Forestry seemed gloomier than the last. Charles Mohr, author of many studies on the Piney Woods over several decades, shared Ashe's pessimism: "The exhaustion of the resources of these forests within the near future is inevitable," he wrote in 1897. The foresters offered working plans for large private landowners across the South; they advised and consulted; they wrote voluminously and spoke to gatherings of lumbermen. But they had no real power, no way to practice what they preached. Not surprisingly, recollected Henry Graves in 1908, "advice given from the standpoint of theory and not of experience often does not inspire confidence."[4]

This situation changed at the beginning of the new century with the establishment of the first national forests. The Forest Reserve Act of 1891 had authorized the president to create national forest preserves, reversing the decades-old policy of selling public lands at $1.25 an acre; ten years later 47 million acres of public lands had been designated as preserves. With the collaboration of Gifford Pinchot and his friend Theodore Roosevelt, these preserves were transferred in 1905 from the General Land Office in the Interior Department to the Bureau of Forestry in the Department of Agriculture. Shortly thereafter the agency had a new name, the U.S. Forest Service, and the forest preserves became national forests.

Now the government foresters had another tool with which to change the behavior of landowners and lumbermen. They were no longer merely cheerleaders but players. On their own forests, they could try to fine-tune the management techniques that they had spent previous years only preaching about. Sharpened by trial and error, scientific forest management practices could then be demonstrated to private landowners, and good forest management would reign over the landscape.

One of the earliest of the new national forests was the Choctawhatchee National Forest, created by President Theodore Roosevelt in 1908. Located in the Florida Panhandle, between the Perdido River on the Alabama border and the Choctawhatchee River, the Choctawhatchee National Forest was transferred to the War Department during World War II and renamed Eglin Air Force Base. When Inman F. Eldredge, the first forest supervisor, and A. B. Recknagel, assistant district forester, arrived at the Choctawhatchee, they beheld a broad, gently rolling, and sandy plain. In places, the sand was almost sixty feet deep, yet on it grew an almost pure forest of longleaf pine. Small numbers of slash pines occupied the wetter sites near the creeks and swamps. Charles Mohr had surveyed the area in the 1890s and discovered that it contained most of the untimbered longleaf left in Florida. Poor soil and limited transportation hindered agriculture and lumbering activities. Lumbermen had culled the biggest trees along the water courses between 1880 and 1900, but the interior of the forest was still intact. Most trees were of advanced age, averaging between 150 and 300 years old, but they grew so slowly in the poor soil that a tree ten inches in diameter at breast height (dbh) — the minimum size for turpentining — was already 100 years old. In comparison, a ten-inch dbh tree on better sites in South Carolina and Texas were 65 to 70 years old. A fourteen-and-one-half-inch tree at Choctawhatchee was 200 years old; a fourteen-inch tree in Tyler County, Texas, was only a century old.[5]

Cattle and hogs foraged on these poor lands, and turpentiners hacked pine trees in the long growing season, sometimes within the cooling influence of the Gulf of Mexico. Twenty-six turpentine companies worked within the boundaries of the forest or on private lands adjacent to it, shipping their products by barge to Pensacola. The soils were so poor that Eldredge and Recknagel decided that Choctawhatchee National Forest would be better off demonstrating turpentining practices than timber management practices. Their management plan was fairly simple: They would turpentine the virgin trees for fifteen years, using the most conservative methods then known, then clear-cut the trees for sawtimber and crossties, leaving two to four seed trees for regeneration. Though aiming at natural reproduction, they planned to experiment with plantations of longleaf pine seedlings and slash pine seedlings, and even with a couple of exotic species, notably maritime pine. *Pinus maritinus* had successfully regenerated the sandy wastes of Les Landes, near Bordeaux, and it was now the mainstay of the French turpentine industry. If it could grow at Choctawhatchee, the government would have a superior turpentine producer, one that might be even more profitable than longleaf.[6]

At Choctawhatchee, Eldredge and Recknagel's management plan almost immediately encountered problems. In line with Forest Service policy, the foresters had directed that all fires were to be suppressed, but they quickly changed their minds after witnessing what happened to the forests as a result. "Three years of actual administration showed that not only would complete fire protection be prohibitively expensive," Recknagel wrote in 1913, "but that it would be unnecessary and dangerous. To keep out fire absolutely over the whole Forest is not only impossible without an extravagant expenditure, but would also serve no useful purpose. The woods would become 'rough' with brush until they were a perfect fire trap; a fire occurring after several years of absolute protection would mean a holocaust, destroying large trees which are now immune." He and Eldredge resumed the folk tradition of annual controlled burning, while raking around the turpentined trees and protecting areas in which the young seedlings were growing. In 1920 and again in 1924, fires were used to prepare seed beds for an anticipated seedfall, but the practice was stopped in 1927 as the Forest Service stepped up its fire-suppression program.[7]

The decades of the 1910s and 1920s weeded out the ranks of turpentine companies at Choctawhatchee. The old specter of overproduction had caused turpentine prices once again to crash regionwide; only five turpentine companies in the national forest survived by 1931, and two of them were tottering. The forest increased the amount of timber cutting in the 1920s to offset the decline in turpentine revenues. Between 1920 and 1930, 10,000 acres of Choctawhatchee longleaf were clear-cut with one to three trees per acre left behind as seed trees.

By 1931 a disturbing trend had become apparent in the cutover areas: the longleaf was not reproducing. The plantations had failed, the exotic maritime pine seedlings succeeding no better than the native longleaf pine seedlings. Worse, the seed crops of 1913 and 1920 fell short of expectations. Ten years after the big seed crop of 1920, only a few survivors were visible, most of them still straggling in the grass stage.

By this time, there were new foresters aboard who unhappily judged the effects of the previous twenty years of management. Summarized forester E. V. Roberts: "The silvicultural system best suited to the longleaf pine type on the Choctawhatchee is yet to be determined. Whatever the cause, it is obvious that we have failed to satisfactorily restock the cut-over areas on the Forest, and to this extent have failed in our responsibility as Forest Managers.

The future success of our plans will depend on the solution of this problem." In the 1931 management plan, he recommended increasing the number of seed trees left on each acre from four to six to see if that would solve the reproduction problem. (The policy was reversed in 1937, when loggers complained that it reduced timber quantity for sale.) He hinted at abandoning natural regeneration entirely, suggesting that wholesale clear-cutting and replanting might be the answer, although the bad results of earlier plantations surely must have given him pause.[8]

Longleaf pine was proving to be a gigantic headache for forest managers, as Choctawhatchee's foresters had discovered. No matter what they did, they just couldn't coax longleaf back onto the cutover lands, at least in the numbers necessary for a profitable operation. The foresters' inability to regenerate longleaf undermined their sense of mission and identity. If you couldn't leave a regenerated forest behind after harvesting, you were no better than a logger, as Fernow had suggested. Scientific forestry's only measure of success was its ability to grow new forests. However confidently foresters had marched off to battle at the dawn of the twentieth century, their silvicultural weapons were failing in the Piney Woods.

———

An even greater problem presented itself—time. Time has always been a challenge for forest managers, and it was even more so in management's early years than it is now. Fernow pointed to the dilemma as early as 1890. "The main difference between forestry and other productive industries," he said, "is the long period of production. From the time of planting to the time of harvest many decades may pass and a century may not be too long. As long as it is not only wood, but size and quality that is wanted, the factor of time is an important one, for it takes time to produce both size and quality." As if to demonstrate this point, Charles Mohr counted the rings of the merchantable longleaf pine logs in one Alabama sawmill and found that the trees were 150 to 175 years old. "Only slow growth under the severe and hardening conditions involved in the struggle for light in the crowded forests," he wrote, could produce the size and the strength of the longleaf pine timbers then being cut.[9]

Today's foresters can shorten time's requirements by thinning and pruning trees, enabling them to add volume faster. But then foresters pondered over the implications that it was necessary to wait 150 years to reap the benefits of forests. What kinds of incentives would encourage landowners to reforest their lands if neither they nor their children nor possibly their grandchildren could profit from them? It was discouraging, wrote Ralph Clement Bryant of the Yale Forest School in 1909. "It requires eighty-five years from seed for a tree under favorable circumstances to reach twelve inches in diameter, and one hundred and eight to reach a fifteen-inch diameter," he explained. "The expense of reforestation and [fire] protection, coupled with the annual payment of taxes, makes so large an investment, when compounded for eighty five years or more, that a timberland owner has cause to doubt his ability to practice profitable forestry on his holdings; and southern lumbermen will not invest money in the improvement of their forest lands until the profit secured from the manufacture of lumber is greater than it is today."[10]

Another Yale forester, H. H. Chapman, studying longleaf pine in Tyler County, Texas, at about the same time, also recognized the disincentives to reforestation caused by longleaf's slow growth. The states could compel forest owners, through regulations, to act against their own financial interests, but it would be wrong, he said. The states would do better to reform ruinous tax laws or buy the cutover lands from private landowners as forest reserves. Beyond that, he believed, there didn't seem to be much else that could be done.[11]

Increasingly, the pressures grew to clear-cut and abandon land. In the days of railroad logging, lumbermen were in no mood to do anything that would take time or put them at a financial disadvantage. The expense of laying railroad tracks, buying steam skidders, and employing skilled manpower meant that owners were going to clear-cut every stick of merchantable wood on a piece of property whether anyone liked it or not. Foresters could only recommend leaving seed trees behind to reforest. The best seed trees generally were the oldest and best formed, from 100 to 175 years old, but a healthy, well-formed, mature tree represented many hundreds of valuable board feet, which lumbermen were reluctant to abandon. Even if you could be convinced to leave healthy seed trees behind, how many were needed? Louisiana passed a seed tree law in 1920 requiring one seed tree at least eight inches in diameter to be left on every acre of cutover forest land, a completely inadequate provision for longleaf because its heavy seeds fell so close to the tree. H. H. Chapman in 1926 recommended leaving not less than four trees, ideally between thirteen and sixteen inches in diameter, forty to sixty feet tall. Choctawhatchee's managers saw two to four seed trees per acre fail to regenerate the forest and recommended six. None of these efforts worked. Indeed, the seed tree method for regenerating longleaf was doomed to continued failure.[12]

Yet though longleaf pine was not reproducing in the numbers or at the speed desired, other trees were. In the Choctawhatchee National Forest, years of fire suppression had encouraged thickets of scrub oaks to grow up on some longleaf areas and seven thousand acres of sand pine had grown up on former longleaf pinelands by 1930, although neither tree was considered merchantable as a timber tree. In many other places in the South, two fast-growing pine species had already drawn notice as marketable second-growth trees. Loblolly pine and slash pine were taking longleaf's place naturally, over vast areas, and both needed little help in growing. It was these species that William Greeley, chief of the U.S. Forest Service during the 1920s, referred to years later in his autobiography, *Forests and Men.* "The restoration of southern pine as a great source of raw material . . . is the most striking event in the story of American forests."[13]

Like many southern pines, loblolly pine (*Pinus taeda*) was known by a bewildering and sometimes deliberately misleading nomenclature. Commonly marketed as "old field pine" or "North Carolina pine," it was also known as "shortleaf pine," "swamp pine," "yellow pine," "sap pine," "bull pine," "black pine," and "Indian pine." In Texas, it was called "bastard pine" because it was considered a hybrid between shortleaf pine and longleaf pine. In Arkansas, it was even sold as "longleaf pine," a species that did not grow in Arkansas.[14]

The timber of loblolly pine ranged in quality from excellent to practically unusable. Virgin loblolly, growing as occasional specimens among hardwoods on the upper, well-drained portions of swamps, yielded superior lumber that matched and even exceeded the excellent qualities of longleaf. Along the fertile Roanoke River bottomlands in North Carolina, for example, one nineteenth-century traveler found loblollies towering 150 to 170 feet high and measuring 5 feet in diameter. Rafts of large-hearted loblolly timbers 50 to 86 feet long were seen being floated north by way of the Dismal Swamp Canal in 1856. Celebrated for their great size and quality, these gigantic trees were marketed as "rosemary pine" in the nineteenth century and coveted by shipmakers all over the world for masts and structural timbers. Yet relatively few of these existed, whereas virgin longleaf dominated the Coastal Plain. By the end of the nineteenth century the great loblollies were gone.[15]

The loblollies that were left had succeeded virgin longleaf after the logged land had been farmed and abandoned. Compared to longleaf, loblolly seeded frequently and prolifically; it released clouds of seeds every year.

When the longleaf was cut, loblollies growing in adjacent areas quickly seeded in the cutover area. If the seedlings escaped fire for a few years, they would grow up in such dense thickets and so quickly that they shaded out the slow-growing longleaf pine seedlings. Once in possession of the land, loblolly was tenacious. Once deposed, longleaf could never again reclaim its kingdom on its own.

By the end of the nineteenth century W. W. Ashe found 100-year-old second-growth loblolly pine growing over much of North Carolina's former longleaf country, and it was increasingly dominant in northeastern South Carolina. "While 100 years ago the longleaf pine was the characteristic forest tree in the Coastal Plain Region of North Carolina," he wrote, "at present the loblolly pine is the prevailing tree."[16]

South of the Savannah River, loblolly was less common and longleaf was increasingly giving way to another pine species, slash pine (*Pinus elliotii*). Like the loblolly, slash pine, or Cuban pine, as it was sometimes called, originally grew in swampy margins of the longleaf pine forest called "slashes" and along streams and around ponds where the trees were often protected from fire. Like the loblolly, it produced a lot of seed at frequent intervals which germinated quickly and grew fast. Its seedlings grew well in shade. As the virgin longleaf was cut, loblolly and slash pines often seeded in the next forest. Stands of slash sprang up in southern Georgia and Alabama, in Mississippi, Florida, and eastern Louisiana—all areas formerly dominated by longleaf. Traveling west by railroad in the early twentieth century from Savannah to Cordele, Georgia, a distance of 160 miles, one observer found that where longleaf pine was once dominant, from 80 to 90 percent of the new forest coming in was slash pine, occurring in "broken and irregular stands, innumerable scattered groups and small stands, in abandoned fields and unused corners about the farm, and along the fence rows, railroad cuts, and embankments."[17]

By the 1910s Forest Service writers enthusiastically promoted slash pine as a superior second-growth tree, celebrating it as "the heaviest, hardest, and strongest coniferous wood grown in the United States." It grew amazingly fast. After six months some slash pine seedlings stood seventeen inches tall, whereas longleaf pine might take ten years to equal that height. Forester Wilbur Mattoon reported that he had found ten-year-old slash pine thirty to thirty-five feet tall and a seventeen-year-old stand that was sixty feet tall and eleven inches in diameter at breast height. Slash also produced more and better quality gum for turpentine. "Intrinsically it is a better tree than longleaf," Mattoon said.[18]

Sawmill operators were among the first to recognize the value of superior specimens of slash and loblolly; even by the end of the nineteenth century they were distinguishing less and less among the southern pines. Longleaf was still considered the superior tree, yet most heart-rich southern pines were classed together and sold as "yellow pine." As a result, the standards were sinking fast. Where once a merchantable log was defined as all heart-wood and at least twelve inches wide, by the time Mohr finished his report on the timber pines of the South in 1897 the standard had declined "nearly to the lowest level, any stick that can be placed on the mill down to 10 inch and 8 inch being fit material." In the nineteenth century second-growth lob-lollies had a poor reputation as lumber. "The wood is sappy and coarse-grained, liable to warp and shrink, and soon decays on exposure," sniffed Moses Ashley Curtis in 1860. Others described loblolly as "the least valuable of the pines" and "the poorest of the southern yellow pines." Now, loblolly and slash pines were eagerly accepted as substitutes for longleaf. The intro-duction of the dry kiln made the sappy loblolly acceptable for a number of purposes, and creosoting made the timber more durable. In foreign mar-kets, the demand for the cheaper loblolly timber was growing. "There was a time when it might have been worth one's while to enter into an argument to demonstrate the value of North Carolina [loblolly] pine as a building ma-terial and for finish of all kinds, but that day has long since passed," wrote the *American Lumberman* in 1907. "The question before the eastern market is not 'What shall we do with it?' but 'How could we do without it?' "[19]

It didn't take foresters or landowners long to see the value in the dense, fast-growing forests of slash or loblolly. One turpentiner in the late 1920s went so far as to deliberately introduce hogs onto his land to root up long-leaf seedlings so he could plant slash pine. Foresters who were still wedded to longleaf tried to compensate by promoting the idea of a "two industry forest." By integrating turpentining into their management plans, a practice that would have horrified earlier foresters, landowners could gain income from both gum and timber on a regular schedule, beginning as early as age twenty or twenty-five. Rotations could be shortened. Treated with a preser-vative, twenty-five- or thirty-year-old longleaf pine timber could be used to make fence posts, small poles, crossties, barrels, crates, and fuel wood.[20]

But these were stopgap measures, too little, too late. By 1930 another in-dustry was just beginning to make its presence felt in the South, one that would revolutionize timber management and end whatever tenuous hold on the region longleaf pine still had. Timber growers were finding that they could cut slash and loblollies as young as ten years old for a market for which timber quality was unimportant. At these early ages, natural longleaf pine was an infant just making its first youthful spurts out of the grass stage, vastly outmatched against jackrabbits like loblolly and slash pines. To the challenge posed by the pulp and paper industry, longleaf had no reply.

—————

The industrial process of using wood pulp to make paper dates essentially from the late 1860s. Previously, a number of different substances had been tried to produce pulp, including, as David Smith enumerates in his *History of Papermaking in the United States, 1691–1969*, straw, cornstalks, bamboo, manila hemp, cultivated hemp, white hemp of Haiti, Indian hemp, cotton, acacia, Spanish broom, silk weed, tobacco, sorghum, pine needles, hops, jute, flax, mulberry, yucca, salt hay, bagasse (the refuse from sugar making), cattails, spider webs, cotton seed hulls, horse manure, Swedish moss, and forest leaves. But in the decades following the Civil War, the processes of pulping wood for papermaking were invented and fine-tuned. Most paper is made from wood fibers pulped by mechanical or chemical means, or by a combination of both. Giant grindstones pulverize logs to make a rough pulp called groundwood; caustic chemicals break down the bonds between the lignin, the glue that holds the fibers together, and the wood fibers. The fibers are washed, beaten, bleached, poured through screens, and then dried to make white paper.

At first, spruces grown in New York and New England produced most of the wood pulp used in American paper mills, but as these forests were logged out, more and more pulpwood had to be imported. By 1930 more than half of the pulp used in American mills was imported, and paper manufacturers were glancing avidly at the second-growth pine forests of the South. Some experiments in papermaking had already occurred in the South in the late nineteenth century, but old-growth southern pines, the kinds that lumbermen were cutting, had a certain notoriety among papermakers. Their problem was one of their natural strengths — their heavy resin content. Whether longleaf, shortleaf, loblolly, or slash, most old-growth pines are thoroughly saturated with resins, especially in their heartwood. Heartwood logs provided the structural strength demanded in the construction trades, but their resin gummed up pulpmaking machinery. Experiments on longleaf pine by the U.S. Forest Service's Forest Products Laboratory in Madison, Wisconsin, proved that a chemical process involving the use of sulfuric acids could dissolve these resins so that a strong, coarse, brown paper called kraft paper could be made from longleaf pine. Southern mills eventually went on to dominate this market. Later experiments by the Forest Service proved the feasibility of bleaching processes that could result in good quality white paper from southern pine. But early on, the specter of southern pines' heavy resin content discouraged mill owners and made them hesitant to invest in locations in the South. By 1911 there were only five paper mills in the South — two in North Carolina and one each in South Carolina, Georgia, and Texas. Only a couple of these were pulping native pines.[21]

Everything changed in the 1920s. A host of industrial boosters that included respected Forest Service foresters like Austin Cary began to flood the pulp and paper industry journals with articles saluting the superior growing capacity of southern pines, especially second-growth stock. What a difference a decade made! For four decades, foresters had obsessed about the expected demise of the virgin longleaf pine forest, repeating their gloomy predictions like a mantra. But by the 1920s, as the last of the old-growth disappeared into the sawmills and as forests of loblolly and slash pine succeeded the longleaf on cutover lands, others were looking at these weedy, scrubby forests and seeing the foundations of a vital new industry. One of these was Charles Herty. Almost thirty years before, in his first appearance in the longleaf pine saga, Herty had made turpentining friendlier for the longleaf pine forest through his campaign for cupping rather than boxing trees. His second entrance, in the theater of pulp and paper, was to alter the timber economy of the South and diminish the role that longleaf pine would play in it.[22]

Herty was a visionary and an industrial genius. Already distinguished in his field in the late 1920s as a retired chemistry professor from the University of North Carolina, ex-president of the American Chemical Society, and former editor of the *Journal of Industrial and Engineering Chemistry*, Herty had returned to Georgia and taken on a new project — proving that the ordinary second-growth trees then proliferating in every woodlot in the South could be the source of high-quality newsprint. He was a publicist of unexcelled abilities, frequently taking the podium before newspaper editors, pulp and paper executives, and audiences of every kind, North and South, to trumpet the news of southern pine. He was such an effective spokesman that enthusiastic newspaper reporters gave him sole credit for discovering how to make white paper from southern pines, ignoring the critical contributions of others. This offended some, who suspected Herty of grandiosity, but the misunderstandings probably didn't upset Herty too much. Haunted by the South's perennial economic problems, Herty had directed his life toward converting the region's natural assets — chiefly its vast pine timberlands —

into commercial possibilities that would enable its people to stand on their own. Any publicity, even sensationalized misinformation, advanced his mission of enticing northern paper mills to the South, a purpose he pursued with an intensity bordering on evangelical zeal.[23]

What he had discovered, in a series of experiments beginning in the late twenties, was that green pine trees twenty years and younger contained little resin. "I have tested slash pine, and I tell you, no matter what others may say, that there is no resin in it," he informed Georgia newspaper editors in 1930. It was the older trees whose heavy resin content were gumming up the grinders, not the younger ones.[24]

In 1932 he set about to develop the industrial processes that were necessary to convert the slash pines and loblollies then springing up in every abandoned field and cutover woodlot in the South into a creamy white pulpwood for newsprint and white paper. By then he had been named research chemist with the Pulp and Paper Division of the Georgia Department of Forestry and Geological Development, stationed at Savannah. The position, and the laboratory and pulping equipment that came with it, was funded by the state of Georgia and a variety of foundations and corporations that were mesmerized by his vision that "from the great open spaces of the South there is going to be turned into the wealth of the Nation a cellulose fiber thoroughly suitable for paper production, ample to supply not only this country's needs but also to build up a great export business." By 1933 he and his research team had successfully created wood pulp from young loblollies and shipped it to a Canadian mill where it was turned into good-quality newsprint.[25]

It was like waking up one morning and discovering that your grass cuttings were worth their weight in gold. It meant that the fast-growing "weed trees," loblolly and slash pines, were no longer the "commonest, worthlessest variety of pine," as one newspaper editor put it, but the foundation for "the greatest development in both agriculture and industry this section has visioned." Realizing they could harvest pulpwood in little more than ten years, rather than waiting for decades as they had been taught, many forward-thinking forest owners saw that forest conservation was now a matter of financial self-interest. "The South is slowly developing an appreciation of its great stores of wealth in forest resources, and is awakening to the necessity of perpetuating them through scientific methods of reforestation and timber use," editorialized the *Manufacturers Record* in 1932. The lure of growing and harvesting pulpwood several times in a single lifetime had finally driven the mass of southern forest owners to take their first serious steps toward reforestation and conservation.[26]

Slash pine boosterism was afoot in the South. In 1935, to celebrate the Slash Pine Forest Festival, the November 11, 1935, edition of the *Savannah Morning News* was printed entirely on paper from Georgia slash pine. Herty contributed a commemorative letter to the edition, touting the virtues of slash pine for its "rapid growth, its high yield of the finest oleoresin for naval stores, and its natural shedding of its lower limbs with the consequent production of logs free of knots and therefore best suited for high-grade pulpwood and for timber." In the same edition was an enthusiastic report on an experiment in which 66,915 slash pine seedlings had been transplanted onto seventy-three previously unproductive acres of woodland, no doubt a euphemism for cutover longleaf pineland. Eighty-one percent of the seedlings survived. The landowner expected to produce income from a twenty-six-year rotation of pulpwood, turpentine, and sawtimber. He would complete an entire forest rotation in just a quarter of a century.

This was the unending cycle of growth and harvest that foresters like B. E. Fernow had described four decades before. It was this vision of an unending stream of forest products that fueled the efforts to get longleaf pine to regenerate, yet, ironically, the vision was being fulfilled by eliminating longleaf from the landscape and substituting pines that were quicker to the lash of human wants. The promise of quick growth keyed the movement of pulp and paper mills into the South in the 1940s and 1950s. In 1925 the pulp and paper industry owned barely 500,000 acres in the South; by 1940 twenty pulp and paper mills were in construction in the South, and timber companies were buying up millions of acres in land swaps reminiscent of the 1880s. By 1942 pulp and paper companies owned 4.7 million acres. Whereas lumber was the dominant forest product in the 1930s, by the 1950s it was pulp.[27]

In the mid-1930s a forest survey had revealed an immense quantity of second-growth pines, but most of this treasure lay in private hands, and timber executives wondered how responsive the owners would be to forest management. The question of timber supply worried them, making them hesitant to move South at first, but reforestation programs beginning in the 1930s eased their fears. Between 1930 and 1945, 2 billion pine seedlings were planted on 2 million acres of cutover southern land, much of it by the Civilian Conservation Corps. Under the Soil Bank Program in the 1950s, billions of loblolly and slash pine seedlings were planted throughout the South—1.7 million acres were planted in pine in 1959 alone. Between 1938 and 1963 forest land acreage in the South increased by 7 million acres, but between 1955 and 1965 longleaf declined by 5 million acres.[28]

As William Greeley said, the reforestation of the South was a major conservation achievement, reducing erosion, protecting rivers and streams, and improving water quality. And Herty's discovery of the economic potential of young southern pines had redeemed the original promise of forestry: to create a permanent forest that would deliver a steady income to the landowner from a sustainable yield of usable products harvested within a reasonably short time. In so doing he had turned Fernow's expectations on their head. Fernow and others had put their faith in the cleansing self-discipline of time to chasten human desire. *As long as it is not only wood, but size and quality that is wanted, the factor of time is an important one, for it takes time to produce both size and quality*, Fernow had said. But by midcentury, Fernow's noble ideal was out of date. With the emergence of pulp and paper markets, timber size and timber quality were no longer the desired values for forests in the American South. Time had been nearly vanquished as a factor in forest management. So had longleaf pine in the process.

ELEVEN

Health, Quail, and Fire

That the exhalation of the pine is directly healing to the diseased lung is no longer doubted. The teaching of Hippocrates, twenty-five hundred years ago on this subject, is still being taught by observing physicians of the present day.
— Dr. Thomas Spalding Hopkins (1888)

Only You Can Prevent Forest Fires.
— Smokey Bear (1944)

The February morning was overcast, but the setting at Millpond Plantation was enchanting nonetheless. There were eight of us in the party. The dog handler and his assistant rode ahead on horseback, two hunters walked behind us, and the others—two hunters, the driver, and myself—lurched about in a mule-drawn hunting wagon that also carried a caged cargo of dogs. Rolling across the landscape on rubber tires, the wagon looked like a cross between a small circus wagon and a pickup truck. From my bouncing perch I could see the longleaf forest dipping and rising, over the hills and down, the dark poles marching off into the distance over a pale winter's growth of wiregrass.

Millpond is directly south of Thomasville, Georgia. If you headed southeast on Georgia Highway 122 out of Thomasville, you'd be passing Millpond property for quite a while before you reached Metcalf. And if you took U.S. 319 to Tallahassee, Florida, you'd ride by a melodious collection of Millpond's sister plantations — Greenwood, Pebble Hill, Sinkola, Elsoma, Dekele, Foshalee, Cherokee, Horseshoe, Mandalay, Welaunee. The plantations are managed traditionally for one purpose: to provide opportunities for their owners and guests to enjoy the pleasures of a winter quail hunt. For the last one hundred years, presidents, governors, movie stars, sports figures, and European royalty have hunted here, not to mention the industrialists and capitalists who created the plantations and who were powerful men and women in their own right.

I was there mostly to soak up the atmosphere of a legendary Thomasville Quail Hunt. The hunt had hardly been legendary so far, more like what my friends in North Carolina encounter as they walk the stubble fields of winter: one covey struck, two shots fired, one bird down, and several uncertain, possibly false, points, the dogs' tails drooping visibly as the hunters approached, their heads turning away, as if lacking confidence in their discovery. Then my host Walter Sedgwick, next to me on the wagon, said, "Point!" and I braced myself as the wagon jolted to a halt. Over the mules' flicking ears I could see two fresh pointers holding staunch on a perfect point, one slightly behind and honoring the first, both rigid in anticipation. The dog handler in his red windbreaker had dismounted and was now in a semicrouch, pointing to a position where the two hunters would get the best shots. The hunters stopped, their guns at the ready.

It was almost a scene out of a painting by Herb Booth or Bob Abbett, a landscape of tall pine trees, golden grasses, chestnut mules, nineteenth-century wagon, dismounted horses, pointing dogs, and somewhere in the stilled grasses a covey of silent quail. "Quail Hunt at Thomasville," reads the title. Every quail hunter in the Southeast has a scene like this in mind when he sets foot into the autumn cornfields, and Thomasville and its plantations have had a lot to do with casting the model. Quail hunters hope that when they go to heaven, they will step right into such a scene. The season will be eternal, the company will be convivial, and the birds will fly right.

Blam! Blam! These birds flew right, one of them anyway, for it practically exploded under the full load of Peter's second shot. There was a flurry of congratulatory voices and Peter, a rancher from northern California, looked pleased, these being his first shots of the day. Everybody seemed pleased — Tupper, the photographer from California; Hacker, the businessman from Nashville; and especially Walter, who had wanted to provide his guests a good hunt. Each of them had come a long way to walk these shooting grounds and engage in this enterprise called plantation quail hunting. As they talked, the wagon driver turned the wagon about and the dog handler picked up the downed bird and, just for a moment, held it to the soft, wet, snuffling noses of the pointers before tossing it into the game box.

For a moment I thought about that bird and the significance of Peter's skillful shot. The few ounces of flesh on the bones of the quail would have hardly fed a cat, yet it had driven the carefully orchestrated hunt and its entire enterprise of artifice and tradition that we had been engaged in that morning. I was struck by the disparity between the bird itself — a six-ounce package of feather, bone, and flash — and the enormous edifice of wealth, ritual, and desire called the Thomasville Quail Hunt that had been erected upon it. It was a disproportion almost whimsical.

It struck me that plantation-style quail hunting in Thomasville was another product of the longleaf pine forest. At the beginning of the twentieth century, longleaf had disappeared in many places, but not in Thomasville. The forests remained a backdrop for pleasant excursions first among wealthy invalids and then among wealthy bird hunters. The invalids were lured to Thomasville, as to other southern places rich in longleaf pine, by the belief that something in the pines was beneficial to their health. After a few seasons, some of them discovered, if not a miracle cure, then something nearly as elusive, the sporting prey of wing shooters everywhere — the diminutive bobwhite quail. It was not a large jump from Thomasville the health resort to Thomasville the sporting paradise, and so the quail plantation was born.

Yet from the search for these two ephemeral products, health and quail, sprang a durable quest among the plantation owners to maintain their pine forests rather than exploiting and devastating them. Out of their quest came an appreciation of the importance of fire in managing longleaf pine, a discovery that influences the restoration of longleaf pine in our own time.

Today Thomasville is a small city of about 25,000 people just twelve miles from the Florida border, but in the 1870s it was starting its run as one of the most fashionable resorts in America. The little promotional booklet published by the *Thomasville Times* in 1888, *Thomasville (Among the Pines)*, shows how the town's origins as a health resort and then as a sportsman's paradise sprang full tilt from the newspaper's copywriters and the town's boosters. Like other communities in the South, Thomasville was engaged in a competitive enterprise to lure the new wealthy classes of the North to the town. The lure the town's boosters extended was based partly on wishful thinking, partly on medical fantasy, and partly on desperation.

The latter was easiest to understand. Before the Civil War, the region consisted of large cotton plantations that shipped their crops by mule team to Tallahassee and then to the port of St. Marks on the Gulf Coast. It was a roundabout and expensive route. Just before the outbreak of hostilities, the Atlantic and Gulf Railroad had connected the region to Savannah, a link that promised to bring the county's goods to port more quickly and end southern Georgia's economic isolation. But after the slave-based economy ended, landowners saw no alternative but to divide their plantations into smaller farms operated by tenants, many of them former slaves. There were 611 farms in 1870 and 1,588 a decade later; by the end of the century the number had doubled again, to 3,183. Low cotton prices, the increased cost of fertilizers, and a lack of capital in the fallen South only increased the desperate straits of landowners and landless farmworkers.[1]

These were the social and economic conditions of the region when the first of the Yankee consumptives detrained in the little town of Thomasville in the 1870s. The first handful must have been a curious sight, handkerchiefs pressed to their lips, cheeks hollow but sometimes rosy with a strange hue. What were these sad-looking invalids doing in the Piney Woods so far from home? Like Ponce de Leon three centuries before, they were seeking a miracle, and when these tuberculosis sufferers looked around and saw all the pine trees, they thought they had found it.

Tuberculosis was once the most dreaded disease in the world. It's hard to imagine that now, with our imaginations so conditioned by the ravages of AIDS, yet in the nineteenth century it was tuberculosis that people feared the most, not smallpox or plague or yellow fever or malaria, killers that all of them were. In the two centuries between 1800 and the present, tuberculosis may have cut down over a billion people. As late as 1930, it put 90,000 into their graves in the United States alone. No wonder it has been called "the greatest killer in history."[2]

The medical profession was helpless to confront this disease. Until 1882, when its germ nature was finally discovered, physicians thought that tuberculosis might be hereditary. Even after the communicable nature of the disease was identified, it would be decades before antibiotics were developed to control it. The contagious nature of tuberculosis made the teeming and unsanitary nineteenth-century eastern cities breeding grounds for the infection. Consumption seemed to arise out of the "horrible toil and turmoil of every-day life in the city," as a nineteenth-century book for invalids put it. At the time, the medical profession was still convinced that diseases and other physical disorders were caused by bad air or "miasmata," the noxious exhalations of decomposing plants or animals. The fogs and mists associated with bottomlands, swamps, marshes, and other low-lying places in the landscape were reputed to be heavy with these poisons, explaining why people living near rivers and along the coast were so susceptible to yellow fever, malaria, cholera, pneumonia, and a host of other fevers and agues. If tuberculosis was caused by damp soil and a lack of sunlight and pure air, it followed that the best remedy for it and just about any fever or lingering illness was a change of climate, someplace where the winter was temperate and the weather was dry, someplace where the invalid could secure daily exercise and plenty of sunlight and fresh air.[3]

To find it, wealthy nineteenth-century consumptives traveled to such far-flung locations as the French Riviera, the Adirondacks of New York, and the rustic outdoor camps of Colorado, Arizona, and California.

And they journeyed to the South, too. There were many disagreements in the medical literature of the day concerning the best southern health re-sort for the invalid. Some doctors who entered these lists seemed to be local boosters first and physicians second. Eastern Florida—St. Augustine and Jacksonville—had its enthusiastic backers, mostly because of the sea breezes and warm climate. Yet others warned against such places. "Experience has amply proved," said one physician, "that a mixture of sea and land air, such as exists on all our maritime situations, is unfavorable to delicate lungs; and especially where there is [tuberculosis], or even a predisposition to it." The coast itself was suspect to these critics because of the dangerous effluvia and the "damp fogs" from the marshes and nearby swamps.[4]

Some doctors began to send their patients to locations in the inland South, accessible by the network of railroads that had been restored to service after the war. One of the places they ended up in was Thomasville. The town was remarkably well connected. By 1883 a web of rail lines joined Thomasville not only to Atlanta, Savannah, Tallahassee, and Jacksonville, but also to such northern and midwestern capitals as New York City, Philadelphia, Cleve-land, Pittsburgh, Cincinnati, and Chicago. There were claims that a "fast mail train" could transport an invalid from New York City to Thomasville in thirty hours.

Taking advantage of these connections, the first touring invalids put up at Thomasville as early as 1870. "Some of the federal troops must have come through here in the winter and talked about the mild winter climate," Tom Hill, director of the Thomas County Historical Society and Museum, told me. "That was pretty soon after the war, when feelings were probably still pretty raw. You have to remember that Thomas County sent 1,600 men to the war and 500 never came back." Thomasville was already soliciting tourists to the region. It had two hotels, but in 1874 the first major tourist hotel, the Mitchell House, was built. In the same year Thomasville native Dr. Thomas Spalding Hopkins gave a speech to the Georgia Medical Asso-ciation in which he claimed: "I doubt if there is on the globe any region of country, of the same extent, more exempt from all diseases of the res-piratory organs." Thomasville had a high elevation and was dry—puddles disappeared just hours after the heaviest rainfall, he explained. The nearest lake was eighteen miles away, the nearest river four miles distant. The sea and its dangerous fogs lay far away. The temperatures were mild, with only a few days below freezing the entire winter. "I have often been surprised at the rapid improvement in my consumptive cases, after removal from the sea-board into that region. I have never seen a case of Pulmonary Tuberculosis in that section of country that could be attributed to climatic influence." If he were an invalid, seeking a winter resort, he said, he'd select Thomasville. The Medical Association unanimously endorsed Dr. Hopkins's opinion and resolved to publicize it throughout the country.[5]

Northerners began to arrive in droves. By 1883 five hotels served the winter crowds and fashionable homes were being built. The winter population shot up to nearly 15,000 people. "Of course, you had to remember, too, that only the super rich were coming," said Hill. "Mr. John Q wasn't here; he was working his six-day work week. Those who were coming ranged from upper middle to filthy rich."

But along with its good climate, railroad connections, and hotels, Thomasville had something that distinguished itself from many other health resorts in the South—the magnificent forests of longleaf pine that surrounded the town. "That the exhalation of the pine is directly healing to the diseased lung is no longer doubted," wrote Dr. Hopkins to a New York doctor. The power of the pines was a staple theme in the promotional writing of the enterprising *Thomasville Times*, which in 1888 published a booklet containing not only Dr. Hopkins's testimonial but those of eleven others as well.[6]

The notion that breathing the air conditioned by pine trees benefited consumptives was already thousands of years old. Hippocrates had reportedly sent patients to Egypt to breathe in the air of the pine forests there, and spirits of turpentine had been used for millennia in healing capacities. The long association of good health with pine trees and with the products of the pine was part of folk medical wisdom among the common people, too. When German traveler Johann David Schoepf visited the South in the 1780s, he noted the sickliness of the people living along the Virginia and Carolina coast and their "pale, decayed, and prematurely old look." By contrast, those who lived in the interior, in the deeper and drier forests, enjoyed a relatively better state of health. "The people themselves are apt to ascribe their better condition of health to the beneficent effect of the pitch and tar odors they are almost constantly inhaling, and they set particular store by the volatile, balsamic exhalations from their pine-woods," he wrote. Schoepf could see the problems with this analysis, pointing out that where swamps adjoined the pine forests, all the tar and pitch in the world couldn't prevent the inhabitants from dying of fever and disease. The two deadliest fevers—malaria and yellow fever—were infectious diseases carried by mosquitoes. Pine forests seemed healthier because there was less standing water on the piney ridges and thus fewer mosquitoes.[7]

The prevailing wisdom drew tens of thousands of northern invalids to scores of Piney Woods health resorts that were sprouting up throughout the Southeast. The Hygeia Hotel, in Citronelle, Alabama, for example, ran an ad in the *Mobile Daily Register* in 1883 pointing out that the town "is noted for the salubrity of its climate, pure water, dry sandy soil, turpentine distilleries, etc., being situated in the pine woods." In the mid-1880s John T. Patrick issued his first brochure on the new health resort of Southern Pines, North Carolina, with the slogan "Consumptives Come to Southern Pines" printed in large letters. A local doctor testified to the beneficial aspects of life in the Sandhills: "Where the long leaf pine exists ozone is generated largely and it has been demonstrated that persons suffering with throat and pulmonary diseases are much benefited when living in an atmosphere impregnated with this gas."[8]

Through the 1880s and into the 1890s Thomasville became a celebrated rural retreat where the social season became just as important as the "cure." Grand hotels were built to accommodate the increasing numbers of guests and their servants who added 7,000 to 10,000 to the town's population each winter. The Piney Woods Hotel, built in the rustic Catskills manner and completed in 1886, had steam heat, gas lighting, electric bells, and an imposing facade that stretched for four hundred feet. "A forest of yellow pine was consumed in its construction," enthused one reporter. It could accommodate three hundred people, most of whom stayed for weeks, months, or the entire season, from December through May. At its peak Thomasville had twelve hotels and twenty-five boardinghouses. The guests promenaded through Paradise Park, a twenty-five-acre pine-studded woodland in the middle of town, enjoyed touring celebrities such as John Philip Sousa and Buffalo Bill, and went for long drives by carriage and horseback through the Piney Woods just outside the town.[9]

Thomasville's run as a fashionable health resort wilted in the aftermath of Robert Koch's findings in Berlin in 1882 that tuberculosis was not caused by bad air or bad genes, but by infection. It took a while before the implications caught on, but when people understood that the disease was highly contagious, the poor consumptives were bundled out of town and discouraged from ever returning. In 1901, when the now *Thomasville Times-Enterprise* published another edition of *Thomasville (Among the Pines)*, it glossed over the town's former identity as a health resort for consumptives. Governor William McKinley, a notable visitor who was soon to be elected president, wrote a testimonial for the publication, but he couched Thomasville's charms in different terms. The town was a "pleasure resort" now, and its main attractions where those of the "outdoor life."

Resorts elsewhere were doing likewise, recasting themselves as winter resorts first, health resorts somewhere down the line. When James W. Tufts issued a promotional brochure about his new Sandhills town of Pinehurst, North Carolina, and mailed it to northern physicians, he stated bluntly: "It is not intended to be a sanitarium." His son Richard later told an interviewer: "My father once told me that when Pinehurst was founded, tuberculosis was generally thought to be an inherited disease. As soon as its contagious character became clearly established, however, Pinehurst veered from its original attitude and as a policy determined to avoid becoming a refuge for its sufferers." When Pinehurst lots were sold in the first two decades of the twentieth century, they contained explicit prohibitions against their purchase by anyone afflicted by consumption or tuberculosis.[10]

Pinehurst went on to develop a gilded national reputation as a golfing resort, but Thomasville followed a different tack. Once its salubrious Piney Woods environment had been demystified, Thomasville lost most of its allure. Aided by Henry Flagler's railroads, the seaside mystique of Florida's resorts now began to seduce the nation's imagination, siphoning off many of the invalids and growing numbers of healthy tourists who had once sought out southern Georgia. One by one, Thomasville's hotels closed or burned and were not rebuilt. Early in the twentieth century the town's resort era was over.

But Thomasville did not fold up its tent and fall back into postbellum depression. Among the social set who had been traveling to the town since the 1880s were wealthy northern sportsmen who had discovered a couple of notable things about the region. One was that the land was cheap and the natives were desperate to sell. The other was that it abounded in bobwhite quail. The result was a buying spree by northerners of cotton plantations in the area that began in the 1880s and lasted for more than a half century, creating one of the greatest concentrations of hunt clubs in the Southeast.

Northerners who had sought Thomasville for the healing balsamic influences of the pines now acquired the pines so they could hunt beneath them. The local property owners were glad to oblige them. "A Yankee is worth two bales of cotton and is twice as easy to pick," went a saying of the time.[11]

By the 1920s several hundred thousand acres of longleaf pine forests in the Thomasville and Tallahassee areas were owned by people who had built expensive country homes and lived there for only a few months of the year hunting quail. These islands of upper-class recreation sheltered pine forests that would never be violated by an industrial turpentiner or a cut-and-run logger. Their main purpose was to produce enough quail to be shot, and they were cared for like a prize heifer or a stud horse.

Despite this cheerful attention, Thomasville's hunting plantations were in trouble by the early twenties. After several decades of remarkable shooting, the seemingly ideal checkerboard habitat of old field lands, virgin forests, and second-growth forests was producing fewer and fewer quail. The land-owners believed they were following the most up-to-date prescriptions for managing their forests. After all, under the tutelage of the U.S. Forest Service, they had stopped the "primitive" practice of burning their woods in the late winter, but the slide continued. It was a worry to the landowners, and a frustration to their guests who counted on a Thomasville hunt each winter. A day of so-so hunting was one thing, but several years of poor hunting were quite another. Without quail, the plantations were almost worthless — the timber had hardly any value, given the overproduction that afflicted the timber industry at the time. A hunting plantation without quail was an expensive and unnecessary hobby.

On April 25, 1923, concerned about the health of the quail populations in southern Georgia, a number of wealthy plantation owners met at the Links Club in New York City. The owners proposed a field investigation into the life history of quail and the reasons for its current decline. Financing the quail investigation was no problem. They quickly put together a private fund and used their clout to seek the cooperation of the chief of the U.S. Biological Survey, E. W. Nelson. One of their number even proposed a field leader for the Cooperative Quail Study Investigation, as it would be known. It was Herbert Stoddard.[12]

Born in Rockford, Illinois, Stoddard grew up in the pine woods near Orlando, Florida, where "everything was pretty much as God made it: a fringe of magnificent and picturesque cypress on the lake margin, and then virgin stands of longleaf pine as far as one could see, arching on and on to infinity," as he wrote in his memoirs many years later. When he was twenty-one, he was hired as a taxidermist by the Milwaukee Public Museum; with his reputation growing, he eventually moved on to the Field Museum of

Natural History in Chicago to prepare exhibits of birds. He joined the army and sailed to France, where he looked for birds and examined the turpentine pines in the forests of Les Landes. After the war he was back in Milwaukee working for the museum again, enriching his bird lore with field studies, and making acquaintances in the new area of bird banding.[13]

Stoddard was thirty-five years old, mostly self-taught and happy with his work in Milwaukee, but the invitation to head the cooperative study in Thomasville was difficult to resist. "I realized that the quail investigation would be a pioneer effort in a comparatively new field," he wrote. "A new profession, wildlife management, was being born." He accepted the job. Nelson charged Stoddard with finding methods that would increase the quail populations. This was no mere academic exercise. "The subscribers to the fund hold their lands in that region mainly for the quail shooting they get and naturally desire a practical outcome to the investigation," he candidly wrote to Stoddard.[14]

What Stoddard found in the South of early 1924 was a landscape mired in confusion and controversy. The main issue was not why the quail were declining, but why longleaf pine was not reproducing. The Forest Service thought it knew why: it was the fault of the free-ranging hogs and the woods-burning practices of the people.

There was no doubt that the huge hog populations that had ranged freely throughout the South for centuries had had deadly effects on the woodlands. A hog ate just about anything, and it was especially fond of the large, nutritious seeds of longleaf pine. A hog could also eat hundreds of seedlings a day. In spring and summer it would root up four- or five-year-old seedlings, eating the thickened taproot and stripping the lateral roots of their bark. A single hog, it was said, could eliminate seedlings from 160 acres of pine forest. Rooting by hogs also loosened the soil and dried it out, creating conditions for hotter and more destructive fires.[15]

Ecologist Cecil Frost has made an impressive analysis of the impact of hogs on longleaf pine. When he added the breeding capacity of the hog to its ability to consume hundreds of seedlings per day, Frost found that hogs' effects across the landscape were multiplied incredibly. "With 10,000 to 40,000 hogs on open range in every settled county in the longleaf region," he writes, ". . . all that would be required to eliminate reproduction would be for a drove of hogs to happen upon a regenerating plot once every three or four years."[16]

Keeping hogs out of pine forests was one solution. In one experiment at Urania, Louisiana, Henry Hardtner of the Urania Lumber Company planted longleaf seeds and seedlings on several plots of cutover land. Two plots were fenced in, the others were left unfenced. After five years, the fenced plots had an average of 6,440 longleaf seedlings on each acre, while the unfenced plots had only 8 per acre. But fencing in forests was considered a hostile act by small farmers and herders who thought of the open-range traditions of the South as a right. It wasn't until the mid-twentieth century that most southern states passed livestock laws that ended the open range.[17]

Even more than hogs, however, it was fire that horrified the new foresters with the U.S. Forest Service. Fire threatened the very basis of the young forestry movement. "As long as there is any considerable risk from fire," wrote chief forester Henry Graves of the Forest Service in 1910, "forest owners have little incentive to make provision for natural reproduction, to plant trees, to make improvement cuttings, or do other work looking to continued forest production." The successful practice of forestry in the United States, he argued, had to be based on protection from forest fires.[18]

From the foresters' point of view, this was a completely understandable argument. The increase in logging activities in the white pine forests of New England, the Lake states, and the western states had provided the fuel for a series of lethal conflagrations. In 1871 the worst forest fire in American history killed 1,500 people in Peshtigo, Wisconsin, and in the same year fires scorched 2.5 million acres in Michigan. Ten years later another wildfire in Michigan ripped through 1 million acres and killed 160 people. In 1894 several million acres were consumed in Wisconsin. In 1902 three-quarters of a million acres of forest were consumed by fire in Washington and Oregon, killing 35 people. In 1910 crown fires burned over 5 million acres of national forests in Montana and Idaho. Many of these fires had spread from fires traditionally used to clear agricultural plots, and in the southern pine forests, careless turpentining operations and slash-littered logging sites created piles of flammable fuel that set off destructive conflagrations.[19]

Complicating the young foresters' attitudes toward fires was their northern training and European bias. The Forest Service at the time consisted mostly of northerners trained in northern schools, principally Yale. These foresters had picked up a European horror of fire from their continental teachers. The last wilderness forests of Europe, of course, had been cut centuries before, and what forests there were grew like hothouse plants, protected and watched over until harvested. Many first-generation American foresters in the late nineteenth century had been Europeans, with Germans such as B. E. Fernow, Charles Mohr, Carl Schurz, and Carl Schenck greatly influencing the young American forestry community. They brought an admirable conservation philosophy to a country that believed in the inexhaustibility of its resources, but they fitted all ecosystems with a Procrustean, one-size-fits-all fire policy.[20]

By the 1920s state and federal foresters were attempting to eradicate woodsburning with every weapon at their disposal. The Clarke-McNary Act of 1924 provided funds to states for fire-control programs and eventually was used to withhold funds from southern states that allowed traditional woodsburning practices. In 1928 the Southern Forestry Congress and the American Forestry Association joined with state foresters in Georgia, Florida, and Mississippi to create the Dixie Crusaders, a group of evangelical young people who brought the message of fire exclusion to villages and hamlets throughout the longleaf pine region. Propaganda films were shown, posters distributed, and lectures given. An estimated 2 million men, women, and children saw these shows; some of them had never seen a movie before. Scenes showing the corpses of deer killed in the hot fires of the Lake states suggested the destructive effects of fire on wildlife. Lecturers declared that fires promoted ticks and even malaria. "Everybody Loses When Timber Burns," was one of the slogans of these crusading fire haters.[21]

For many people in the Piney Woods, this commotion over their annual fires was nonsense. Southern herders had burned the woods annually for two centuries or more. The people believed that fire kept down the insects, greened up the grass for cattle, produced fruiting berries for wildlife, reduced flammable fuels, brought game to browse, and performed other useful things. To them, woodsburning was natural and even necessary. But outsiders experienced culture shock when they saw the people move into the woods with lighted torches to touch off fires that burned over thousands of acres. The experience of one northern landowner in 1894 in southern Georgia was probably typical. "On the last day of the quail season," he recollected, "our head Negro 'made a narration' to the tenants giving them permission to 'put out the fire,' which, the buildings and fences having already been raked around they promptly did, 'putting the fire' to everything else that would burn. That night, on every hand, lines of flames crept or raced across fields, flickered through pine woods, here and there flaring high over the heavier clumps of weeds, accompanied by cracklings of brush, bangs like pistol shots, and clouds of eye- and nose-stinging smoke." It was the custom of the people, one person told him, and it took place every spring as long as his great-grandfather could remember.[22]

Fire historian Stephen Pyne describes the different attitudes toward fire as cultural differences—the old culture of Indian woods lore that had been adopted by the European and African people of the Piney Woods for several centuries versus the "scientific" culture of the new forestry and agricultural experts. Writing about the attitudes of these cadres of young professionals, Pyne says: "The real tyranny of technology transfer in this instance was that a self-proclaimed science demanded the repudiation of frontier folkways. Centuries of practical experience painfully and empirically acquired by American settlers in a range of fire regimes were abruptly sacrificed, only to be rediscovered later."[23]

Stoddard was one of several inside and outside the Forest Service who played key roles in this rediscovery, providing evidence that the decline of quail and longleaf pine were intertwined and that both were related to the Forest Service's hostility to fire. He spent five years on fieldwork before his landmark conclusions were published in 1931 as *The Bobwhite Quail: Its Habits, Preservation, and Increase.* He examined the life history of the quail, their behavior and habits, their feeding preferences, their movements, mortality, population fluctuations, predators, parasites, and popular beliefs about the quail. He examined quail crops to find out what they were eating, performed experimental plantings, observed the natural processes of the seasons. It was immediately clear to him that current agricultural methods were eliminating quail habitats. The intensive cotton culture in the region had undoubtedly depleted soil fertility, reducing its ability to grow the seed-bearing weeds and legumes that fed quail. Modern methods of "clean" agriculture were removing thicket cover and concentrating quail in habitat islands that attracted predators. Many forests in the region were too densely overgrown to produce quail foods. Quail thrived, Stoddard said, where there was a diversity of habitats — open woodlands, weedy fields, cultivated fields, thickets, and scattered grassy areas.[24]

The Bobwhite Quail became an instant classic in the infant field of wildlife management, but it was almost never published. Stoddard's conclusions had the enthusiastic backing of the Biological Survey and the Division of Food Habits Research, under whose auspices the Cooperative Quail Study Investigation had been initiated. But they were opposed by Forest Service personnel who objected strenuously to the chapter entitled "The Use and Abuse of Fire on Southern Quail Preserves," even though Stoddard had made his observations "as mild as my conscience would permit." Because of Forest Service objections, Stoddard rewrote the chapter five times and, finally, knowing he had the support of his wealthy backers, threatened to resign and write an account that would not be watered down. The book was then cleared for publication.[25]

Stoddard's chapter on fire was only one of twenty chapters in the book, but it was the most controversial. The first sentence of this chapter was like throwing a gauntlet down in front of the experts. "The Bobwhite of the Southeastern United States," Stoddard began, "was undoubtedly evolved in an environment that was always subject to occasional burning over." He had suspected that fire played an important role in regulating quail populations almost as soon as he started his studies, and his initial impressions were confirmed by them: Quail were declining at least partly because fire had been excluded from the landscape. "[Fire] may well be the most important single factor in determining what animal and vegetable life will thrive in many areas," he wrote. Frequent patchy, though not annual, fires were critical because they maintained the open habitat in which quail thrived. Without fire, the dominant wiregrass and broomsedge grass shaded out the legumes and other quail foods, whereas a February fire enabled legumes and perennials to seed in abundantly. Slow-creeping fires helped quail get at the seeds that they craved but did not harm the forest. Weak scratchers, quail "could starve in a land of plenty" if their food was buried under a mat of grasses and pine needles.[26]

Stoddard's conclusions for quail management were relatively simple: Keep pinelands open; eliminate dense thickets of shrubs and other shady vegetation through periodic fires; use night fires in winter, not spring or summer; use plowed fire lanes to contain fires and carefully supervise their use.

His recommendations were wise, at least compared to the policy of fire exclusion then in vogue, but in some respects they were primitive. Plowed fire lanes twenty to forty feet wide, which he recommended to control the spread of fire, are destructive. In many cases they are placed in the wet ecotones where many rare and endangered plants reside and where fires will put themselves out naturally, if skillfully handled. The winter fires Stoddard preferred to spring fires often do not kill back the hardwood thickets. Stoddard's message, moreover, was marked by caution and equivocation, evidence of his fierce struggle with the Forest Service over nuances of phrasing. He described fire as a "convenient, though not always a vitally necessary, tool" that was "capable of doing vastly more harm than good if not intelligently handled." Indeed, he wrote, his recommendations were aimed only at the management of quail preserves, not properties interested in pursuing timber or agriculture where quail could only be considered as "supplementary." His counsels "should not be used to embarrass the forester in his attempts to protect forest growth over the region at large."[27]

Yet, there was no denying Stoddard's subversive message: Fire in the pine-lands would bring back the quail.

Would it also bring back the pines themselves? There were many who thought so. Referring much later to his standoff with the Forest Service, Stoddard would immodestly call himself "the most outspoken advocate of controlled burning in coastal-plain pine forests." He was certainly outspoken, but he had lots of company. For forty years or more, reasoned opinion regarding the importance of fire to the longleaf pine forest had abounded inside and outside the professional forestry community. Though the Forest Service seemed to speak with one voice sternly denouncing fire, the agency was actually a churning cauldron of contradictory opinions. At Choctaw-hatchee National Forest in Florida, assistant forester A. B. Recknagel was only one of many Forest Service personnel in the South who believed that the folk practice of burning the woods frequently to prevent larger fires was a good one and that fire, correctly employed, would aid in the regeneration of the forest.[28]

Outside the Forest Service, there were many scientists who saw the critical importance of fire in the Piney Woods. As early as 1888, a speaker at the American Forestry Congress in Atlanta had pointed out that in areas that had been protected from fire, hardwood trees had invaded. "Viewed from a forestry standpoint we believe the total abolition of forest fires in the South would mean the annihilation of her grand lumbering pineries," she concluded. Botanist Roland Harper reached the same conclusion in the first decade of the new century. Having examined hundreds of forests in almost every state in which longleaf grew, he wrote: "If it were possible to prevent forest fires absolutely the long-leaf pine — our most useful tree — would soon become extinct."[29]

Another pugnacious champion of fire was H. H. Chapman, the Yale professor who would have very close and critical ties with the Forest Service. He argued that southern pines were different from northern pines and that fire-exclusion policies designed for the latter would be disastrous in the longleaf pine forests of the South. "It is the right policy for Northern States, where fires can and should be absolutely prevented," he wrote in 1912. "But there is abundant evidence that the attempt to keep fire entirely out of southern pine lands might finally result in complete destruction of the forests." From 1917 to the 1940s Chapman and his students from Yale researched longleaf pinelands near Urania, Louisiana, from which fire had been excluded since 1915. They showed that fire suppression caused more hardwoods to spring up, resulting in the germination of fewer longleaf seedlings. Chapman, like

Stoddard, believed that the common practice of annual woodsburning was harmful to longleaf seedlings. He suggested that seedlings could be protected from fire for a few years to make sure they survived. His experiments showed that frequent fires would eliminate competing pines and hardwoods and also rid the longleaf pines of brown-spot needle disease, which had spread widely since the fire-suppression campaign had begun. Accusing government foresters of ignorance, Chapman would be a thorn in the Forest Service's side for nearly a half century.[30]

While Chapman was busy in Louisiana and Stoddard in Georgia, S. W. Greene, a researcher with the Bureau of Animal Industry (BAI) in Mississippi, was proving that cattle put on more weight when feeding on burned forestlands than lands protected from fire. Fires not only eliminated competitive vegetation, he discovered, but they also encouraged the growth of legumes and grasses that fattened cows. "Forest owners may yet turn to the use of fire to fight fire and get fire insurance for the cost of a match by knowing how and when to use a match as the natives of the southern piney woods have known for generations," he wrote. The Piney Woods people had a truer understanding of fire than the foresters. "Their conclusions were not set down as scientific treatises but were passed along under the more common name of woodcraft and the lore of the woods told them when, why and how to use fire."[31]

Greene also had a hard time getting his findings published. The BAI was a sister agency of the Forest Service within the Department of Agriculture, and though Greene's data had a direct bearing on Forest Service policies, the service wanted to avoid confusing the public and encouraging indiscriminate woodsburning. Agency protocol required a consistent point of view nationwide regarding the evil of fire, despite growing evidence to the contrary. Suppression of data and self-censorship were the result during this uncertain period. When the director of the Southern Forest Experiment Station summed up the Forest Service research that conclusively supported periodic fires, one of his colleagues cautioned him: "I doubt the feasibility of telling the public about [controlled burning in longleaf pine forests] at this time. It may be wrongly interpreted." Greene finally was able to publish his findings in *American Forests*, the journal of the American Forestry Association. Titled "The Forest That Fire Made," his article was preceded by a lengthy and unusual editorial disclaimer that included this statement: "[Greene's] conclusions that the Longleaf forests of the South are the result of long years of grass fires and that continued fires are essential to the perpetuation of the species as a type will come as a startling and revolutionary theory to readers schooled to the belief that fire in any form is the arch enemy of forests and forestry."[32]

Greene's paper appeared in 1931, the same year that Stoddard published the results of his quail investigation. In 1935 the issue of "Forest Fire Control in the Coastal Plains Section of the South" was on the agenda of the Society of American Foresters' annual convention in Washington, D.C. Chapman organized the session. One after another, the speakers agreed on a single conclusion: Longleaf pine reproduction would not follow from fire exclusion as they had once predicted. "In the longleaf region the quantity and inflammability of fuels, and hence the fire hazard, increase under continuous fire protection," admitted E. L. Demmon, director of the Southern Forest Experiment Station. Winter fires could reduce the fuel loads, he said, prepare the seed beds, help control brown-spot needle disease, and improve pasture and game conditions, the very things that Harper, Greene, and a host of Forest Service researchers had been saying for over thirty years. For the first time, the agency was publicly admitting what it had been privately suppressing for decades.[33]

Officially, there was now a new attitude on the part of the Forest Service regarding fire in the South; unofficially, the old antifire attitudes prevailed for decades more. The result was an ambiguous and confusing fire policy. At one point late in the 1930s, when the issue was all but over and the old ideas of the settlers vindicated, the Forest Service hired a psychologist to investigate why the people of the pines burned the woods. His answer: Their beliefs that fire benefited the pine forest were "the defensive beliefs of a disadvantaged culture group." In 1943 the Forest Service finally sanctioned prescribed burning programs in the South, and in five years nearly a million acres were burned, most of them in Florida. But a year later the agency introduced a Disneyesque cartoon character named Smokey Bear who sternly admonished the American people that "Only You Can Prevent Forest Fires." It was a spectacularly successful public relations program, but one that undermined public education about the necessity of prescribed fire for decades to come.[34]

In retrospect, it is easy to canonize the Piney Woods people and condemn the condescending bureaucrats from the North. Yet fire in longleaf pine forests looked very different to professional foresters at the turn of the twentieth century than it does to us a century later. At that time, industrial turpentiners, moving fast across the landscape, were in effect painting the forest with gasoline by slashing trees to make the flammable resin run. They often killed trees within a few years by cutting multiple boxes and chipping deeply into the cambium. Cut-and-run loggers added their debris to the volatile mix. The result was a forest with more dead and diseased trees, more blowdowns, and more fuel on the ground than natural forests might have contained. What horrified foresters observed in the Piney Woods was a tinderbox waiting to explode.

Annual fires also were harmful, eliminating as much as 90 to 95 percent of a longleaf stand's potential reproduction. And the people were hardly blameless in their use of fire. They often used fire carelessly and maliciously. Woodsburning could have numerous motives, including jealousy, revenge, and pure meanness. "If people got upset with you, they'd set your woods on fire," said one timber company owner in Virginia. The intensive silvicultural practices of twentieth-century timber companies often drew the ire of hunters, especially when these companies destroyed acorn-producing scrub oaks in the process of planting and managing for pine. One early-twentieth-century jingle summed up the sentiments of a lot of southern folks of the time:

You've got the money,
We've got the time,
You deaden the hardwoods
And we'll burn the pine.[35]

Yet however malicious the impulse, a fire may not have been as destructive as it seemed to northern eyes. And however laudable the motive behind the Forest Service's policy of fire suppression, it only succeeded in increasing fuel levels many times over what was natural. "The fact is, the high fuel levels in our forests today were created by the very policies that were designed to prevent catastrophic fires in the first place," wildlife biologist Jay Carter told me. "That's something that's lost to the general public at large and probably lost to the forestry people as well." Fire suppression also allowed forests to grow denser and to crowd out the longleaf and its associated plant and wildlife species. It was disastrous for the southern forest, and what's left of today's longleaf pine woodlands are still paying for this error. Said one retired Forest Service researcher, trying to explain the service's hostile attitude toward fire for most of the twentieth century: "Smokey Bear could not distinguish between a fire that warmed a house and one that burned it down."

TWELVE

Fools for Longleaf

I'd be a fool to plant longleaf. Why would I plant longleaf when I can thin
loblolly and slash pine in fifteen years and make some real money?
—From a conversation overheard in a Florida café

In 1935 only about 20 million acres remained out of the original 92-million-
acre longleaf pine forest. Of that, perhaps three-fifths was second-growth
longleaf that had seeded in accidentally after the biggest trees had been
taken, another third was bare, cutover land that had been skinned clean, and
the rest, hidden away in patches here and there, ignored or coddled by their
owners, was the remnant old-growth. By 1955, 12 million acres were left; ten
years later there were only 7 million acres; and by 1985, after waves of spectacular conversions of longleaf pine forestland to plantations of loblolly and
slash pines, to row crops and parking lots, longleaf pine in the South could
be found on only 3.8 million acres.[1]

Forest Service administrators were so pessimistic about the future of
longleaf in southern national forests that in the 1970s they proposed closing
the Escambia Experimental Forest, in Alabama, which had been testing
methods of growing and regenerating longleaf since 1947. Researcher Bill
Boyer was told, "Longleaf pine has no future in the Southern forest. It's a
dead species."[2]

Yet in 1986 the Forest Service publicly proclaimed its commitment to reverse the century-long decline of the species. What had happened?

Several things had happened. One occurred in the 1980s when environmental groups used the Endangered Species Act to sue the Forest Service
on behalf of the endangered red-cockaded woodpecker. In losing many of
these court cases, the Forest Service was forced to adjust the management
of its longleaf forests to protect the red-cockadeds and other endangered
and threatened organisms. Out of this court-ordered husbandry came many
advances, not the least of which was a reversal of the Forest Service's long-

standing practice of converting longleaf pine sites to loblolly and slash pine. Now southern national forests had to put longleaf back. That story will be told in the following chapter.

Perhaps more important, however, was the Forest Service's realization that the problem of regenerating longleaf pine had been solved. Forest Service researchers had discovered not only how to regenerate longleaf pine naturally, from existing seed sources, but also artificially, from planted seedlings. In the world of the longleaf pine, these were revolutionary accomplishments.

For decades, the goal of regenerating longleaf pine had seemed almost a will-o'-the-wisp, a mirage that continually evaporated in the field. As a result, longleaf pine had developed a reputation as a cranky and unreliable species. In 1946 Forest Service researcher W. G. Wahlenberg, in his authoritative study *Longleaf Pine*, could write with some understatement, "Deliberate regeneration of longleaf pine has been rarely accomplished. In fact, the reproduction of this species has been so irregular and uncertain, and its natural controls so imperfectly understood, that successes and failures have been difficult to explain." Well into the 1950s and 1960s failures in regenerating longleaf persisted, driving more landowners to loblolly and slash. Pulp and paper companies led the way, followed by many southern national forests, all of them establishing vast plantations of loblolly pine.[3]

If foresters were to keep longleaf as an option for landowners, they had to be able to offer an alternative to clear-cutting and seed-tree cutting, two regeneration methods that had failed with longleaf pine. Clear-cutting and replanting had been common harvest and regeneration methods used in the South since about 1950. They worked fine with loblolly but failed to regenerate longleaf because no seed sources were left behind to regenerate the stand naturally and because no one knew how to plant longleaf seedlings. Seed-tree cutting seemed to offer an improvement, but foresters consistently underestimated how many trees were necessary to seed a single acre. If you left a few loblolly pines per acre after a harvest, you would easily regenerate the entire tract. Why wouldn't a few longleaf do the same?

As foresters discovered, loblolly was the wrong model. With its annual production of masses of light seeds, a few loblollies easily produce enough seed to regenerate a harvested stand. But longleaf produces seed on an irregular basis and much less of it. The longleaf's seeds are much larger than a loblolly's, and they are so heavy that they fall closer to the tree than loblolly seeds, severely handicapping the longleaf's ability to seed in a large clear-cut.

The development of an alternative to clear-cutting and seed-tree cutting for longleaf happened almost by accident at the 3,000-acre Escambia Experimental Forest in Brewton, Alabama. Escambia was one of several research stations in the Southeast where U.S. Forest Service researchers plugged away at understanding the needs of longleaf pine. In the early 1950s Tom Croker and his Forest Service colleagues were examining 40-acre plots that had been clear-cut each year from 1948 to 1951, three plots per year, twelve plots in all. On each acre, eight seed trees had been left behind, a conventional seed-tree strategy for regeneration. What Croker noticed was that the three 40-acre stands that had been cut in 1948 were flourishing with vigorous young longleaf pine seedlings, while the other nine, harvested in succeeding years, were not.[4]

What had happened was not a result of the eight seed trees left on each acre. In 1947, the year before the first cut, longleaf trees in the central Gulf Coast region produced one of the two greatest bumper seed crops ever recorded (the other one occurred in 1996), with cone counts of about 150 per mature tree. Many seedlings had germinated on each of the twelve plots. When three of the plots had been cut in 1948, the seedlings on those plots grew quickly in the absence of the competing trees. On the plots cut just two years after seedfall, the seed crop had failed, overwhelmed by animal predators and competing trees and vegetation. The key to the successful natural regeneration of longleaf, then, wasn't just the occurrence of a big seedfall, but the removal of the mature trees very soon after seedfall. Even two years after seedfall was too late.

This was close to what foresters had known as a shelterwood cut, a European silvicultural method that had been used in other parts of the United States but never in southern pines. A shelterwood cut is similar to a seed-tree cut in that some trees are left behind after the harvest to act as seed sources but then are removed after the seedfalls to allow the seedlings to grow. In longleaf pine forests, that was easier said than done. The problems were twofold: How many trees were sufficient to regenerate a stand? And given the irregular seeding habits of the tree, how did a manager know when to expect an adequate seed crop?

The questions were daunting, but the method seemed so promising that Croker and his young associate Bill Boyer set to work trying to come up with the answers. The number of seed trees was critical. Too many, and they would compete with and restrain seedling growth or create so much needle litter that they would fuel fires hot enough to kill the young seedlings. Too few trees, of course, would not adequately seed in the stand and create so little fuel for fires that hardwoods would spring up, dooming the seedlings. Over decades of experimentation, the Escambia researchers tested anywhere from ten to fifty trees per acre before deciding that about twenty-five to thirty trees per acre at thirteen inches diameter at breast height would provide sufficient seed to regenerate the plot.

Once they remove all but thirty trees per acre (the number is approximate, depending on the size of the trees), foresters use binoculars to monitor the cone development on those trees each spring. It may take several years before a sizable cone crop develops, but eventually one does come along. A good crop will produce about 750 to 1,000 cones per acre, each containing about fifty to sixty seeds. It's important to know when an adequate seed crop is expected because foresters can then plan a spring fire to eliminate the undergrowth and prepare the ground to receive the seed the following fall. If the seeding is successful, the seed trees are removed after a year or two. If not, the trees remain until the seed crop does succeed.[5]

Boyer says that he wants to make sure that 5,000 to 6,000 seedlings are growing on every acre before the overstory trees are removed. That may sound like a lot but it's a bare minimum. Loggers will kill half while removing the thirty adult trees that are left, and perhaps 90 percent of the rest will die in the first year of disease, predation, and fire. About 300 to 600 surviving disease-resistant seedlings per acre will be enough to regenerate the stand.

The development of shelterwood silviculture in longleaf pine has been one of the few silvicultural triumphs in a century of so much disappointment. Understanding the nature of longleaf cone and seed production gave foresters a new predictability over seeding, and by working within this system foresters had a regeneration method that they believed resembled nature's way of growing a new longleaf forest. Today the shelterwood method is reliably regenerating the species on many sites across the South.[6]

The second silvicultural advance was the development of successful techniques to plant longleaf. Some notable success in planting longleaf had been achieved by lumber companies in the 1920s, but planting failures were still common for decades to come and discouraged many from even trying to establish new plantations of longleaf. Whereas loblolly reforestation had been accomplished easily, the same techniques applied to longleaf were disastrous. The two very different species require completely different nursery methods. One forester facetiously told me that about the only way you can fail with loblolly is to plant them upside down, and even then some may live and grow. Not so with longleaf. A longleaf pine seedling has to be bigger when it is planted and handled more delicately.

It wasn't until the 1970s and 1980s that Forest Service foresters working in

Alexandria and Pineville, Louisiana, developed nursery techniques so long-leaf could be more successfully hand- or machine-planted. When it came to planting longleaf, landowners now had two options: They could plant either the less expensive bare-root seedlings or the more expensive seedlings in containerized packages.

With many problems in longleaf regeneration overcome, representatives from the Forest Service, the forest industry, and academe met in 1986 to discuss the longleaf decline. Encouraged by the achievements in longleaf silvicultural technology, they set out to prove that landowners need not fear longleaf any longer. "The time has come, and in fact is overdue, to initiate an 'organized campaign' to get this technology disseminated and transferred to the southern forestry community operating in the longleaf pine range," the group declared. Assuming that knowledge would vanquish decades of frustration about growing longleaf, the group planned a blitzkrieg of conferences, symposia, tours, papers, slide shows, and videos. They encouraged workshops by state chapters of the Society of American Forestry and forestry extension divisions. The goal was "to stop the reduction in longleaf pine acreage by 1990 and experience a net increase by 1995 over 1990 acreage."[7]

In due time many of these efforts were carried out, and they helped to spawn others. Between the late 1980s and the end of the 1990s, scores of state and regional conferences on longleaf pine attracted landowners, foresters, and ecologists. The meetings were often divided between presentations by ecologists and biologists about the role of fire and the species interactions in longleaf ecosystems, and presentations by foresters about natural and artificial regeneration, the use of chemical sprays to reduce competition, and other technical subjects. Foresters and ecologists were often mutually suspicious of each other's motives, yet there were still many gains. In 1988 Fred White, North Carolina's chief forester at the time, candidly admitted to me that southern forestry had boxed itself in with its "myopic obsession with loblolly pine." "The reputation of southern pine as a structural pine was made by longleaf," he said, "and it was made in the period from 1900 to 1950. We have been riding on that reputation ever since." Loblolly was beginning to give the southern lumber market a black eye because the fast growth of the tree produced weak wood. It didn't season well and didn't saw well, and in the construction trades carpenters were turning to more stable Canadian spruce. "This is a price that industry hasn't really reckoned will have to be paid," White observed. He required his state foresters to attend workshops to reeducate and retrain them about the economic viability of longleaf for the small, private landowner. The process of changing minds began first with the foresters, he told me; then the landowners could be brought around. He described a future in which longleaf would be rightfully restored to at least some of its original range.

Yet in 1991 longtime longleaf pine researcher Bill Boyer, analyzing the most recent Forest Survey results of longleaf pine acreage, found a decline in every diameter class from one to fifteen inches, indicating that few young longleaf pine forests were being established. "At present, the outlook for this forest type is not promising," he wrote. By 1995 Forest Service statistics showed longleaf's acreage had dipped once again, to 2.95 million acres. The greatest losses—about 72 percent of the total—were occurring on lands belonging to what foresters call private nonindustrial landowners, the small landowners who owned the bulk of the remaining longleaf.[8]

The Forest Service had underestimated the effort it would take to recruit landowners who had been alienated from longleaf essentially through the service's own attempts earlier in the century. Like the Smokey Bear campaign against forest fires, the loblolly jingoism that federal and state foresters had encouraged for so long proved stubborn and long term. It would take more than a flurry of meetings and a few glossy pamphlets to convince foresters and landowners to grow longleaf pine.

The core of the problem still remained the character and the needs of the private landowner, as Fred White told me. "The sort of landowner that it would take to manage longleaf we don't have now, or at least very many of them—that's one who would plan ahead. This is the problem. Many landowners don't even think about their forests until they need funds, and then they cut. All of the techniques we know now about longleaf require advance planning and commitment prior to cutting. It's the rare landowner that does this."

These landowners were still rare in the 1990s when I traveled the South looking for longleaf, but I found some landowners who were giving longleaf a chance.

One of these true believers lived in southern Mississippi, and one spring day I found myself driving with him across farmland whose rich soils had once nourished some of the biggest and densest longleaf in the South. This was where many of the turn-of-the-century photographs of longleaf pine forests were made, the ones showing what seemed to be monster columns of virgin trees dwarfing a standing man. Just west of the little town of Columbia we crossed the Pearl River, and as I peered through the truck window at the brown water below, I couldn't help but think of the great forests of virgin longleaf that floated these muddy currents on their way to the Gulf and beyond.

Thinking about them gave me a perspective on what I was about to see. Mickey Webb braked and spun the wheel of his pickup. We left the highway and drove for a few minutes down a dirt road, his shock absorbers protesting as we dipped and lurched through potholes. We parked near a gated road and climbed the fence to walk beside a thick, brushy field. Scattered amid the Bahia grass I could see the darker green of sapling longleaf pines about twelve feet high.

"This is about a 600-acre tract," said Webb. "It's had a pretty rough history. Before we cut it recently, it was a scattered mix of loblolly and shortleaf. The previous manager wasn't really managing it; he was just cashing in. We're going to plant it back to longleaf and what I want to show you is a longleaf plantation we put in here about 1991.

"I'm real pleased with it. You just don't know how long we've been trying to grow longleaf here. We started in 1986 with some bareroot stock but that didn't do well at all. Then in 1991 we put in some containerized longleaf seedlings and we've gotten good survival out of them, about 85 percent. We're getting probably three feet of growth a year. Look at that one!"

He pointed to a gangly twenty-footer. "He's putting on what, about five feet a year?"

I nodded and tried to share in his enthusiasm, but the field looked weed-choked and rough to me. It was a couple of minutes before I caught the pride in Webb's voice, and then it came to me. "You just don't understand how long we've been trying to grow longleaf pine here!" he had said. *He was planting longleaf pine.*

Webb was a forest consultant in Columbia who had been on the front lines of longleaf pine restoration in Mississippi for more than twenty years. He'd been the president of the Mississippi Forestry Association and a past editor of "The Consultant," a newsletter for consulting foresters. A devout Christian, he prefaced many of his sentences with "We've been blessed" and didn't hesitate to ask me if he could say an audible blessing when we stopped for a big plate of barbecue deep in the Mississippi countryside.

Webb was also a believer in longleaf pine on private lands. He thought that ecologists and foresters had invested their hopes for the future of longleaf on public lands—national forests, state forests, wildlife refuges, and the like—and had given up on a role for privately owned lands, even though their owners controlled far more acreage. Most of Webb's clients were owners of small to moderate-sized forest properties totaling more than 80,000 acres, and of that amount, only 3,000 acres actually had standing longleaf on it. That meant that if his clients were going to be convinced to grow longleaf, they would have to plant it. He said the biggest challenge of managing smaller properties is to get their owners to reforest their land in the first place, much less reforest with longleaf pine. In many areas throughout the South, landowners treated their 100-acre forest as a bank account, liquidating it when they needed cash and letting Mother Nature grow another forest. In Mississippi, about 100,000 acres a year were cut without reforesting, and for Webb that was where the challenges and the opportunities lay for restoring longleaf pine.

"For many years, none of my clients ever suggested planting longleaf. I had to convince them," he told me. "Most of them just didn't know anything about pines. If you asked them to name a species of pine they'd say 'field pine.' They thought of that as a species. Others have a very negative bias toward longleaf and when I've suggested that they might grow some longleaf, they go ask another forester and he usually tells them, 'Whatever you do, don't let him plant longleaf!'"

He laughed. Webb battled with county foresters all the time to approve state cost-sharing funds for clients' forest plans that included longleaf. In Mississippi, until the 1990s, they were reluctant to approve cost sharing for longleaf and in the past actively discouraged landowners from growing it.

"There are three criticisms of longleaf: it doesn't regenerate well, and when it does it stays in the grass stage too long, and when it finally emerges, it grows too slowly," Webb said. "We've overcome those hurdles in the last fifteen or twenty years. Our biggest problem now is in getting the message out." The message was simple: We know how to grow longleaf pine now.

How do you sell longleaf to landowners? I asked Webb.

"First I tell them they are restoring the ecosystem," he said. "This is an opportunity to practice good forestry and help nature. I tell them that the very best forestry tries to work with nature as closely as possible, and the reason longleaf was the dominant forest type for so long here in Mississippi was because of trial and error over tens of thousands of years.

"Then I tell them that longleaf pine is by far the most resistant of all Southern pines to fire and to every known pest and disease—annosus root rot, fusiform rust, the Southern pine beetle, the pine tip moth. I also argue that they shouldn't have all their eggs in one basket. That's the most effective argument, I've found. These superior loblolly pine seedlings that everyone planted come from just a few genetic stocks, and if some epidemic or tree disease comes in you're finished. You know, if stocks are doing well you invest in bonds just to give yourself some diversity. It's a risk to have everything in one species of tree.

"And then there's the economics of it. I say, look, longleaf makes a superior tree and there's no question in my mind that there's always going to be a demand for high-grade saw timber and poles, if nothing else for rich folks who like pretty veneers. Yeah, longleaf pine is more expensive to grow than loblolly, but not that much more and you get an excellent return."

In recent years, forest economists have taken on the task of demonstrating to landowners that growing longleaf makes money. Several management models have emerged, although none of them disprove the caution that growing longleaf successfully and profitably is a long-term, sustained effort requiring more care, discipline, and self-control from the landowner than any other southern pine. Recommended rotations range from fifty-five years to eighty years and longer, which makes the typical landowner grimace. But within that eighty-year period, longleaf yields a series of products that get more valuable with age and in the long term may well be more lucrative than growing loblolly or slash for pulp.

The secret is time, however, and that's a commodity few landowners respect. Pulpwood and fence posts can be harvested from early thinnings, and high pulp prices in recent years are part of the economic argument that Webb employed. "Now I can tell someone you pay $50 or $100 reforesting his tract in longleaf and I can thin it out in just ten or twelve years, $200 to $250 an acre. We're not talking about thirty or forty years down the road. If you have 100 acres, that's $25,000. That's real money." Poles and sawtimber can be cut at age thirty years and yield increasing value for the next three decades. Poles are used by utility companies for their telephone and electrical wires, and they are the most valuable of longleaf's economic products, selling for almost five times the price of pulpwood. For this reason, most consultants like Webb consider clear-cutting longleaf for pulp a waste of money and advise their clients to wait for the more valuable products later on.

We drove to another property, where Webb showed me yet another way a client can break his heart. It was a broad, grassy field studded with the dark green of early longleaf pine seedlings. Not long ago it had been part of a 250-acre stand of 65-year-old longleaf, but the landowner clear-cut it out of fear that it would attract red-cockaded woodpeckers. Under the laws protecting the endangered bird, landowners were restricted from cutting some of the trees around cavity trees. In this part of Mississippi, where there were so few red-cockadeds, the possibility of them colonizing an old stand of pines was remote. Before he got the jitters, the landowner had been removing timber worth $40,000 a year from his longleaf, and he had more timber in 1993 than he did in 1979 when he began his management program.

"So from a cash flow point of view, there was no need for him to cut the timber," Webb said. "It was just a panic. We've got gopher tortoises here, too, and they're a protected species, but if we keep managing reasonably, gophers will do all right."

At another site, Webb nodded appreciatively at the little seedlings and saplings that were growing, and he talked about using chemical herbicides to rid the soil of vegetative competition so his little seedlings would grow fast. "This is bare-root stock, machine planted," he said as we strolled through a brushy, recently planted field. "It's the exact same age as another nearby tract where we didn't do any chemical treatment. We got a lot better response here than there. I think we'll go chemical from now on."

He stopped suddenly and looked at the ground where some purple berries were growing.

"I love it that the gophers have been eating the dewberries," he said.

As we moseyed through the flat southern Mississippi terrain, I thought about Mickey Webb's management challenges. So completely had longleaf been eliminated from most private lands in Mississippi that Webb had had to build longleaf pine forests from scratch using more intensive management practices than many conservationists were comfortable with. He used machines to plant seedlings and chemical herbicides to control the hardwood competition. He confessed that he once felt guilty doing this. "I came into forestry during the environmental movement of the 1970s," he explained, "and it made an impression on me. I didn't use chemicals at first because I felt they were an environmental problem. But I've learned I've got to be able to control things, so I plant and I use chemicals to knock back the competition. One of the worst things that could happen for longleaf regeneration, I think, would be the loss of chemicals."

Webb used fire as his major hardwood control. He burned in the winter and spring, although he was doing a lot more spring and summer burning now because it tended to control the sweet gums and other hardwoods more efficiently than winter fires. "If I burned in February, I might kill a six-foot-tall sweet gum back to the ground," he said. "Well, by the following August it's four-and-a-half feet tall because it's sprouting from the root—it had all those nutrients stored in the root waiting for the spring leaf-out. So what I like to do is first burn a couple of winter fires to reduce the heavy fuel load that's on the ground. The hardwoods sprout out in spring and use that big reservoir of food stored in the root and then I burn a fire through in April, May or June. There's no question we're getting more complete kill right to the roots."

Webb wanted his trees to get up out of the grass stage as fast as they could, so he didn't employ fire as often as some, a decision he attributed to findings by Forest Service researchers Bill Boyer and Bob Farrar that burning can decrease tree growth, although they caution that this is a price that sometimes must be paid. Consequently, his land was brushier and less open than other longleaf pinelands that were burned more frequently. The ground cover was missing.

As we drove back to Columbia, I thought about Mickey Webb and what he represented in the longleaf pine restoration efforts that I was beginning to see. Economics played a central role in most of the forest plans he put together. "Hey, if we're going to bring millions and millions of dollars into longleaf pine restoration, we've got to be able to show the degree to which it will pay off," he told me. "Folks are willing to take on the additional cost that goes along with managing longleaf if they know they'll get something from it."

Not that all they want is maximum income. In Webb's office at the end of the day, I met 83-year-old landowner Sedgie Griffith who was delighted with the stands of longleaf that Webb was managing for him. When he and his brother bought 250 acres of mostly cut-over land in 1951, only young longleaf were growing on it, he told me. "But those little trees just kept coming." A twister spawned by Hurricane Camille leveled about 80 acres in 1969, but he planted again and bought more land. "We've sold over $600,000 in timber from 400 acres," he said. "I have had the pleasure of seeing five grandchildren pretty well go through college from what that tree farm brought.

"And I'm still planting. It seems strange, you know, to be eighty-three and still be planting trees."

"You must believe in the future, then," I observed.

"I do, I believe in the future," he said in his gentle Mississippi accent. "And I'm convinced that putting longleaf pine on these sites is the best thing. I can remember driving to Sumrall as a boy. There were still good stands of longleaf pine out through that area. My father was up in Mt. Olive and I remember driving there in an open Ford car, driving through those beautiful virgin longleaf stands. You can't imagine how pretty it was down in this country."

I drove northeast from Mobile to Escambia County, Alabama. Only three counties in the South have more than 100,000 acres of longleaf pine and Escambia is one of them. The other two lie just across the Florida line, in the western portion of the Panhandle. The three counties nestle together like pieces of a jigsaw puzzle, giving the region the distinction of having the greatest concentration of contiguous longleaf pine in the country. In the southernmost county, Okaloosa, Eglin Air Force Base (464,000 acres) sprawls across the landscape, while in adjoining Santa Rosa County to the north, the giant Blackwater River State Forest (189,594 acres) is the major landholding and major repository of longleaf. Just across the Alabama line is Escambia County, where Conecuh National Forest resides (84,000 acres). The county is also marked by the presence of large, privately owned properties, vestiges of the huge timber baronies that once held sway over the region. Brewton is the county seat and the home of the Cedar Creek Land and Timber Company, formerly the T. R. Miller Mill Company. Rick Jones was its assistant general manager, and if there was a single reason I visited Brewton, it was probably a remarkable claim that I came across in a paper delivered by Jones at one of the numerous longleaf pine meetings in the Southeast. He had ended his talk by saying, "Good luck! And do not 'close the door' on longleaf; it has helped keep my employer viable for 123 years."

One hundred and twenty-three years of managing longleaf? I wanted to find out how and why.

According to its in-house history, the T. R. Miller Mill Company dates to 1872, when Elisha Downing bought 1,200 acres and a water-powered sawmill on Cedar Creek that had been cutting virgin timber for over thirty years. In 1892 the company built a steam-powered sawmill on Murder Creek, in Brewton, and by 1912 it had amassed about 83,000 acres, which it proceeded to strip of most of its virgin trees by the 1920s. Other large timber companies also operated in the area, but one by one most of them went out of business until T. R. Miller and Alger-Sullivan of Century, Florida, were the only survivors. By the 1950s Alger-Sullivan sold its 400,000 acres to a half-dozen smaller pulp and paper mills that clear-cut the remaining longleaf and planted slash and loblolly pines. T. R. Miller was still growing predominantly longleaf in the mid- to late sixties when one of the controlling families sold a large chunk of the property to the Container Corporation of America. To retain controlling interest in the company, T. R. Miller took a loan, clear-cut thousands of acres of longleaf to pay off the debt, and replanted the land in slash and loblolly. About 80,000 acres of its 200,000-acre holdings remain in longleaf today.

When I joined Jones in his office at the mill, he was almost apologetic about this chapter in the company's history. "Longleaf pine seedlings were just not available," he explained. "That's the reason we're not utilizing all our longleaf pine acres in longleaf pine." Given the advances in longleaf pine regeneration technology, Jones said, the company planned on converting the off-site pines back to longleaf.

In the 1930s one of its forest superintendents used the trunk of his 1929 Chevrolet Coupe as a kiln, filling it with longleaf pinecones, collecting the seeds that shook out, and planting them. This first plantation failed, as did many others. But visits by Yale University forester H. H. Chapman and Austin Cary of the Forest Service led to a program of frequent burning, which, probably more than anything else, saved the longleaf that remains today on T. R. Miller lands. The company is a fierce proponent of frequent burning, setting 20,000 to 30,000 acres ablaze every year.

I asked Jones to explain why the company had stuck with longleaf pine for so long.

"Our sites are suited to it, first of all," he told me. "They're typically sandy soils with some gravel in them. They've got a site index of seventy to seventy-five, something like that." He explained that "site index" refers to the height that you can expect trees to grow on a particular piece of land in fifty years. A site index of seventy is about average for growing longleaf.

"And a big part of Miller's business has been in producing poles, and it has been ever since the 1920s," Jones said. The company operated two sawmills: a pole-finishing mill and a treating plant. Jones said that T. R. Miller made 4 percent of the telephone poles in the United States, and the percentage was increasing.

"Longleaf pine makes good poles, and longleaf pine is pretty much what made T. R. Miller, but, believe me, if it didn't make economic sense, we wouldn't grow it. I guess we feel like, there's a lot of little trees; we're growing bigger trees," Jones said.

The clouds were the color of bruises, and the day carried a humid promise of rain when I left Jones's office with forester Dan Head. As we drove deep into the heart of the old T. R. Miller property, the skies opened and a gray curtain of rain briefly lashed the banks of the tall longleaf lining the road and the wind tossed their tops. At an old section called "Whynot," I wandered through the wet grass and gallberry shrubs, gandering at the large trees. According to company lore, this was the exact spot where T. R. Miller and other officials in 1905 decided to carry logs to the mill by railroad, rather than float them to the mill in the ditches. "Why not?" they asked, and this part of the property had carried the name ever since.

Head explained that different managers even on T. R. Miller/Cedar Creek had slightly different styles. "Say there were some oak trees scattered out there on a tract that we want to cut. Now I may leave some, but someone else may cut all the oak out of it. But as far as managing the longleaf, everyone probably has got the same idea. Try to keep the stand healthy. Use plenty of fire. And when the tree gets to a certain age and there's a big seed fall, be prepared for it—if you want to regenerate naturally, I mean."

Because of its standing longleaf pine, T. R. Miller tried to regenerate longleaf naturally as much as possible, using the shelterwood system developed by Tom Croker and Bill Boyer at Escambia Experimental Forest.

"T. R. Miller is on a sixty-year rotation," Head said. "But we have some leeway. If I see a pretty vigorous stand that's sixty-years old, for example, I may not cut it to a shelterwood even though the age justifies it. If it's a healthy stand, I may just select cut it, take a few trees or a small group of trees.

"Now a paper company, they may be growing loblolly on a fifteen-year rotation, and in some places it's starting to drop to twelve. In other words, every fifteen years they cut it and start all over again. That's a tree farm."

Tree farms are increasing their domination of the Alabama and southern landscapes. In 1952 natural pine stands occupied 72 million acres while planted pines covered 2 million acres in the southern states. By the end of the century, 47 percent of the pine growth in the South was in pine plantations; in Alabama, pine plantations exceeded natural pine forests by more than a million acres.[9]

Thus the sixty-year rotations of a company like T. R. Miller were unusual in a state where the number of acres in pine tree farms had nearly doubled during the 1980s, where one-fifth of the state's total tree mass was loblolly pine, where the U.S. Forest Service predicted that 70 percent of the state's woodlands would be in pine plantations by the year 2030.[10]

At T. R. Miller, then, I was looking at a rarity, even an anachronism—a commercial forest whose owners still believed in growing longleaf. Decades ago, they had answered the question "Why grow longleaf?" and the answer still made sense to them.

When Jim Morgan was growing up in the North Carolina Sandhills, he considered it a boring place and couldn't wait to leave. He set out for Atlanta and stayed there for fifteen years. But a funny thing happened when he returned. The place didn't seem as boring any more.

"I saw it was really special because this Sandhills ecosystem is such a rare thing, just a little strip of land coming out of Georgia and South Carolina, with the really white sand and that ecosystem of longleaf pine and wiregrass, and this handful of animals adapted to it," he said.

Now he took a special interest in the farm, especially the 1,700 acres of longleaf pine growing on the property. He had enrolled Morgan Farms in the Forest Stewardship Program, which helped landowners construct a long-term management plan for their forests with the aid of foresters, wildlife biologists, and soil and water specialists.

"My Dad has a land grant that goes back to 1789," Morgan told me. "There are some trees that survive from that period. We've cored a few that go back two hundred years. It's just been part of the family tradition that you take care of the land and are a steward of the resources." He was happy to see that wildlife had begun to return because of the management plan that he had installed. He'd found wild turkey, fox squirrels, and several coveys of bobwhite quail on his land after years of absence.

Morgan's pinelands also sheltered several colonies of red-cockaded woodpeckers, and unlike most people, he said, he was glad to see them. Many landowners believe that if they manage for older longleaf pine forests, they will only encourage these endangered species and end up with burdensome restrictions on their right to harvest trees. That's the reason why many landowners have clear-cut the longleaf on their land, thus eliminating the possibility that such forests would attract red-cockadeds or other endangered species. Rather than doing that, Morgan enrolled his land in the Safe Harbor Program. This federal program, administered by the U.S. Fish and Wildlife Service, offers participating landowners economic incentives to help protect endangered species. By signing the Safe Harbor agreement, Morgan will not be responsible for any additional birds attracted to his land as a result of managing his forests in ways that protect the woodpecker and encourage its habitat.

"I think there are a lot of people confused about red-cockaded woodpeckers and the Safe Harbor Program," Morgan noted. "It's a great program, but the average landowner still thinks of the birds as an 'infestation.'"

It had not been easy to be a good forest steward, especially in the Sandhills. "Sandhills soil is really an oxymoron," said Morgan. "Longleaf pine trees are what this sand is designed to grow. But you're beating your head against the wall to try to grow just about anything else except maybe tobacco." Yet Morgan and his family had turned what seemed to have been a bad hand into a winner. He was looking to add more acreage to the longleaf pine already growing on his property. The reason? Pine straw.

Pine needles have always been a minor economic product of the longleaf pine for which few uses were found. Indians wove them into baskets, and in Louisiana the Coushatta Indians continue the tradition today for the tourist trade. In the 1880s in Columbus County, North Carolina, a mattress manufacturing plant employed workers to stuff mattresses with needles from felled trees. Sixty years later the upholstery and mattress-making markets were declared a "new use" of the golden needles. Pine straw was also used as a sometime mulch on agricultural lands, a fertilizer for cotton fields, and a stable bedding in many parts of the South.[11]

But in the 1970s and especially in the 1980s, pine straw was rediscovered as a mulch for gardeners and landscapers, and the new market has recast this humble product of the longleaf pine into "brown gold." Longleaf pine straw is longer, has a more attractive color than loblolly pine straw, and can be baled like hay. Pine straw raking can be a year-round enterprise. In the Sandhills an acre of longleaf pine can produce from sixty-five to one hundred bales of pine straw each year, netting a dollar a bale for the landowner or the thief—pine straw poaching is a cottage industry in the Sandhills. Pine straw holds more value than sawtimber. As long as property owners keep trees on their land, they can rake straw and make money year after year. There's no wait.

As a result, in 2002 pine straw was a $50–$55 million dollar-a-year industry in North Carolina, one that was driving what can only be described as longleaf mania. I have talked to landowners who are looking for any scrap of land to plant trees, even a piece as small as five or six acres. This craze for longleaf seems to be the closest contemporary equivalent to the mania for turpentine that swept through the region a century ago. The lure of profits drove that mania, and dollar signs are feeding this one. Given longleaf's persistent reputation as a money loser, I wonder if that isn't an improvement.[12]

"When we made our management plan in 1994, it had not become as clear to us and perhaps to others the enormous cash value of straw," Morgan said. "We had no idea how much money could be made by raking straw." In the 1990s pine straw skyrocketed in value. In the Sandhills, an acre of straw can produce sixty-five to eighty bales of pine straw, and Morgan has gotten quotes of seventy-five dollars an acre for someone to come to his property and rake. In some cases, property owners have even doubled this return.[13]

Trees as young as ten to fifteen years old produce the best straw, according to Morgan. What this means is that with the addition of pine straw to the economic equation, longleaf pine can generate a constant income stream throughout the life of the stand, even from young trees in plantations. If you add to that the value of the timber as it enters the sawtimber and pole stage, it's a money-making combination.

"We've never wanted to just maximize dollars," Morgan pointed out. "We've always wanted to say, we have a beautiful resource here, let's find a way to balance the environment and wildlife and the business part of it. You've got to pay the taxes on the land each year so you have to find a way to make money."

Pine straw raking has been a driving force in maintaining the pine overstory in the Sandhills, Mark Cantrell of the U.S. Fish and Wildlife Service told me. "It has kept a lot of natural pine stands from being converted to loblolly or developed into trailer parks. And now after a couple of years of good pine straw raking, we're seeing folks recognize that they can keep their land in longleaf and actually make some money at it."

For the landowner, the best news is that even a newly planted longleaf pine forest yields pine straw relatively quickly—normally in fifteen years and in some cases in as little as eight. Thereafter it can be raked every year or even several times a year.

Of course, that's where the problems arise. Many ecologists view frequent raking to be little more than a form of mining, especially when it's done by tractor-mounted rakes. Mats of needles play an important natural role in the sandy soils of the Sandhills forests by holding moisture at the surface where the shallow root systems of the pines can get at it. They are sources of nutrients for the trees and other plants, and they provide the fuel necessary to carry the periodic and life-giving fires that ensure a healthy and reproducing forest.

"I've seen places that have been raked twice a year, where they rake it every time a needle hits the ground, and year after year," Cantrell said. "Those are the places where you'd see the herbaceous stuff and the wiregrass being uprooted by the rake tines. You lose your vegetation and you don't get any pine regeneration. After a while that's going to be a losing situation."

Indeed, pine straw raking may well be impoverishing one of the most diverse ecosystems on earth. Ecologists consider the Sandhills one of the centers of biodiversity in the South. "Pine straw raking should not be regarded as a harmless cure for financial problems," warned ecologists Michael Schafale and Alan Weakley. "Extraction of 'brown gold' is accompanied by a long-term degradation of ecological systems." Terry Sharpe, a wildlife biologist with the North Carolina Wildlife Resources Commission, put it bluntly: "You can manage for pine straw and provide everything that red-cockaded woodpeckers need, but you can't manage for pine straw and provide the diverse ground cover with all the necessary components to have a healthy population of grassland wildlife in your pine stand. You have to make a choice of whether to have pine straw or healthy grassland wildlife populations."

Morgan tried not to put all his eggs in a single basket. He cut some timber, and he still farmed some of his acreage, although he was diverting more farmland into longleaf pine plantations. He preferred to rake the young stands more often and the older stands less often, because he knew the value of the needles as fuel for prescribed fires.

"If you read our management plan it basically just says 'Burn!' " he stated. "So we burn everything in a regular planned three-year cycle." After adopting the plan, he and manager Dave Buhler set their prescribed fires mostly in winter to reduce the hardwoods that had grown up in thickets. But now they were doing growing-season burns.

"We think of growing-season burns as more natural—most of the fires happened because of lightning strikes, and that had to have been during the growing season," Morgan explained. "We felt proud of the condition of our forest this year. We had two fires from campers who were camping on our property illegally, but the fires didn't do any damage at all. They just ran along the ground. Had that happened ten years ago, we would have had major trouble."

Morgan cut some timber from his property every year, usually through small-scale modified shelterwood cuts in which some of the mature seed trees were retained and might be removed in ten to fifteen years. "When I first got here I didn't think I would cut anything. A consulting forester took me to another property. We walked all over and didn't see a single quail. Then he took me to a place where they had opened the forest up through a shelterwood cut and there was an explosion of quail."

He believed in forest diversity and preferred having multiaged stands of trees to single-age classes. Though his burning program and selective use of herbicides were thinning out the scrub oak thickets on the property, he didn't want to get rid of all of them. With wildlife as one of his goals, he wanted to keep some high quality mast-producing trees.

One of the motives that Morgan guarded against was greed. "You can't be greedy," he emphasized. "People sometimes want to get all the money and right now, but you can't do that. You really have to take a little off the table each year and continue to invest. You've got to keep burning and planting. It's sustainable over a long period of time. You know, it's a beautiful thing to walk up on a piece of property that is well managed and see and hear wildlife and know that you're still actually making money on the land."

PART FOUR

Ecosystem Restoration

THIRTEEN

Woodpeckers and Forests

If some systematic method be now sought for in the management of long-leaf pine forests, it will be well first to glance again at the natural conditions under which they grow and upon which such a method should be based.
—G. Frederick Schwarz, *The Longleaf Pine in Virgin Forest* (1907)

It had been a real accomplishment, hard-gained, to be able to say that long-leaf could be managed for the valuable products it made without compromising the ability of the forest to regenerate. But by the 1980s newer developments in the society at large were eclipsing these achievements and making them seem almost irrelevant. The Forest Service's timber-oriented forestry was under fire nationwide, and demands were growing for forestry practices that were aimed less at profit and more at ecosystem health and other goals. Now values such as biological diversity, old growth, nongame and endangered species, recreation, and even spiritual restoration were taking their places at the forest management table, and they were changing the way longleaf pine was managed.

The National Forest Management Act of 1976 requires each national forest to develop a land and resource management plan with public participation. By the late 1970s and 1980s these plans and their review and revision process had become convenient targets for environmental groups throughout the country, with the result that many of the plans were bogged down in litigation. The most far-reaching legal challenge occurred in Texas in a case that became a defining moment in the management of longleaf pine on public lands. For more than a decade, several environmental groups had been sniping at the Forest Service over its management of several Texas national forests. The groups based a number of lawsuits on alleged violations of the National Environmental Protection Act, the Wilderness Act, the National Forest Management Act, and the Endangered Species Act. One case, briefly successful, was reversed on appeal (*Texas Committee on Natural Resources v.*

Berglund). In two other cases, the endangered red-cockaded woodpecker and the Endangered Species Act played small roles, but in 1988 the woodpecker stepped fully into the spotlight.

In the case that became known as *Sierra Club et al. v. Lyng et al.* (*Sierra Club v. Lyng*), the Sierra Club, the Wilderness Society, and the Texas Committee on Natural Resources based their lawsuit squarely on the Endangered Species Act and the protections it offered to the endangered red-cockaded woodpecker. They argued that the Forest Service plan for the Texas national forests had violated two provisions of the act—Section 9, which prohibited a "taking" of an endangered species, and Section 7, which specifically enjoined any federal agency from any activities "jeopardizing" an endangered species. Their evidence was statistics gathered by biologist Richard Conner and ecologist Craig Rudolph, both with the Forest Service, that showed spectacular declines in the populations of the red-cockaded woodpecker between 1983 and 1987 in three of Texas's national forests—the Sabine, the Davy Crockett, and the Angelina. If a taking meant "to harass, harm, pursue, hunt, shoot, wound, kill, trap, capture, or collect," the plaintiffs argued, the Forest Service had certainly "harmed" the woodpecker. Moreover, the decline of the species in Texas jeopardized its survival and recovery.[1]

The outcome of the lawsuit changed the way forestry was practiced in longleaf pine forests on public lands in the Southeast. It's difficult to predict how the suit would have been ruled had it not been for the red-cockaded woodpecker and for the increasing sensitivity about the status of endangered species in general. Nevertheless, henceforth the provisions of the Endangered Species Act would be powerful tools in changing forest management principally, though not exclusively, on public lands.

Biologists have been concerned about the red-cockaded woodpecker's status for some time. Some believe that the population has fallen by more than 99 percent from its pre-Columbian levels. What's left of these birds is concentrated in fifteen population centers throughout the Southeast, but only one of them, the Apalachicola National Forest in Florida, has a viable population, meaning that it contains at least 250 groups that are successfully reproducing. For the U.S. Fish and Wildlife Service to consider the species recovered, six of the fifteen populations will have to be viable.[2]

The woodpecker was classified as a federally endangered species in 1970. On many of the older cut-and-run sites, the few old "cull trees" left on each acre had often been sufficient to provide cavities for the birds, but with the growth of intensive forest management few old trees survived the clear-cuts, and the dense young pine plantations offered nothing of value for the woodpecker. Since the 1960s forest rotations in the southern national forests had been shortened until they were no more than sixty years. The removal of the old trees had devastating consequences on the woodpecker, as did decades of fire suppression that enabled dense thickets of scrub oaks and other hardwoods to supplant the original open forest, forcing the woodpeckers to abandon their cavities.[3]

Natural forces were also unkind. On the night of September 21, 1989, Hurricane Hugo slammed ashore near Charleston with winds estimated at 135 miles per hour. It devastated the city of Charleston, practically wiped out the small town of Summerville, and though its wind speed typically slackened as it went inland, it still packed an incredibly destructive punch as far inland as Charlotte, North Carolina, and West Virginia. The eye of the hurricane passed just south of Francis Marion National Forest, which until then had had the second largest population of red-cockaded woodpeckers. Portions of the forest's midsection bore the brunt of some of Hugo's fiercest winds with gusts measured at 145 miles per hour. Particularly hard-hit were the older longleaf pine trees. Surveys immediately after the storm showed an estimated 50 to 60 percent of the sawtimber trees destroyed. On ten thousand acres with longleaf more than fifty years old, about 95 percent of the trees were twisted off or uprooted. In five hours, the storm had destroyed more timber in the state of South Carolina than any other natural disaster in history.[4]

The storm also decimated the red-cockaded woodpecker population. Sixty-three percent, or slightly more than 1,200, of the birds were unaccounted for and presumed killed; more than 1,500 cavity trees in loblolly and longleaf—87 percent of the total—were destroyed; and 70 percent of the foraging habitat for the woodpecker had disappeared, all in a few hours. Only 229 of the original 1,765 cavity trees remained. Before Hugo, the national forest had generally been regarded as having the healthiest and the densest red-cockaded woodpecker populations in the South, but after Hugo the population plummeted to about 239 family groups and fell further during the following winter.[5]

Recognizing the desperate situation that the bird was in across its range had been the easy part for biologists; doing something about it was more difficult and contentious, and ended by polarizing the scientific community. Recovery efforts were mounted as early as 1975. A team of biologists cobbled together a management plan in 1979 that recommended, among other things, thinning longleaf forests to open them up, eliminating hardwoods, and using more prescribed fire. It recommended rotation ages of one hundred years to provide older trees for cavities, something the Forest Service had been reluctant to do because of the impact on its timber goals. The plan was revised in 1985 to take advantage of new knowledge about the life history and habitat needs of the bird, but other biologists attacked it almost immediately. The American Ornithologists' Union concluded that the Forest Service was unwilling to reduce its timber goals sufficiently to make the kind of effort needed to recover the woodpecker.[6]

It was about this time that the Texas environmental groups brought the suit that would be known as *Sierra Club v. Lyng*. (Lyng was Richard E. Lyng, secretary of the Department of Agriculture; others named as defendants included F. Dale Robertson, chief forester of the U.S. Forest Service.) According to the plaintiffs, clear-cutting, seed-tree cutting, and shelterwood cutting were the culprits behind the precipitous decline in the woodpecker populations in the Texas national forests. Known as even-aged management practices, they produce a forest stand consisting predominantly of trees of the same age.

Even-aged management, which had become policy for southern national forests in 1963, is based on the observation that certain tree species do not tolerate competition well and regenerate best in cleared, open stands. The young trees need room to grow. They need plenty of sunlight and room for their root systems to suck up moisture and nutrients. Grasses, shrubs, hardwood trees, and wildflowers rob the young planted trees of root space, nutrients, and light, and so these competitors have to be eliminated or the seedlings will fail. Site preparation is the way to get rid of competition. Site preparation can be accomplished easily with prescribed fire or with other means that include chemical herbicides, disking, or chopping the soil. Often, woody debris is piled into windrows and burned.

Growing in even-aged stands, the young seedlings receive full sunlight and are protected from the depressive effects of older trees. They can be thinned periodically to emulate the natural mortality of younger trees. This kind of forest management can coax trees into their full growth potential in the shortest time.

Of course, time is exactly what is at stake in this notion of "intolerance." The growth of a longleaf pine, after all, can be released at any time during its long life. A young tree can sit in the grass stage for years, even decades, before beginning height growth. And even a mature tree, suppressed beneath a dominant tree, can put on an amazing spurt when the dominant tree falls or is harvested.

Commercial forest managers, however, want to make trees grow faster and more efficiently, and to maximize the economic potential of their forest and create a sustained yield of timber, an even-aged approach makes sense. Imagine a 1,000-acre forest. To manage it according to even-aged principles, the forest or tree farm can be divided into, say, ten 100-acre stands, each of which is assigned a finite life span called a rotation. If the rotation age, or the age at which the oldest stand will be allowed to grow, is one hundred years, then the well-regulated even-aged forest will eventually consist of a 100-acre stand that is ten years old, one that is twenty years old, one that is thirty years old, and so on, all the way to one that is one hundred years old. The stands are cut regularly according to a schedule. If the cutting schedule is ten years, each block advances in age as a cohort, is thinned regularly, and is harvested when it reaches one hundred years. The result is a sustained flow of timber.[7]

Under a scheme like this, trees grow fast, they grow together, they follow your schedule. The forest is fairly easily compartmentalized under an even-aged system, making it simple for work to be scheduled and completed—a prescribed burn here, a thinning there, a harvest cut there. The schedule is dictated by the needs of the grower. Regeneration takes place on schedule, at least theoretically.

This is the industrial model of forest management, and it has made the South one of the wood-growing capitals of the world. For large landowners such as the timber industry and the Forest Service, which manage hundreds of thousands of acres of land, the ability to organize forest management in this way is a major accomplishment. And it's a big advantage if the objective is to grow the most wood in the shortest time and make the most money from your investment.

But can even-aged management satisfy the needs of the red-cockaded woodpecker?

The plaintiffs in *Sierra Club v. Lyng* argued that it would not. Clear-cuts, they said, eliminated foraging or cavity habitat for decades, and they fragmented the woodpeckers' habitat, creating isolated islands of habitat in which the birds found it difficult to forage and find mates. Longer rotations of one hundred years or more would have produced the older trees that the birds needed, but such long rotations were practically nonexistent.[8]

District Court Judge Robert M. Parker Jr. agreed with these arguments.

"The practice of even-aged management has resulted in significant habitat modification," he wrote. "This is not merely a situation where the recovery of the species is impaired by the agency's practices . . . but rather the agency's practices themselves have caused and accelerated the decline in the species."

Timber harvesting and endangered species protection are not easily attainable simultaneously, Parker reflected, yet Congress had eliminated language from the Endangered Species Act that would have allowed federal agencies to implement the act "insofar as is practicable and consistent with the[ir] primary purposes." Instead, the Endangered Species Act imposed a duty on federal agencies not just to do what they could, but to make endangered species protection the highest of priorities. Agencies were to use "all methods and procedures" that would enable their survival.

To this end, Judge Parker issued a permanent injunction against the Forest Service's even-aged management techniques, ordering a number of remedial measures to be taken, primarily the conversion of forest harvesting techniques "from even-aged management to a program of selection or uneven-aged management that preserves 'old growth' pines."

It was an astounding judgment. Uneven-aged management springs from a different conception of the managed forest. Rather than an aggregation of blocks of even-aged trees, a forest managed according to uneven-aged principles consists of trees of many ages spread somewhat randomly throughout the forest. There are no age classes, no rotations. The forest is the entity, not the stand or compartment. Trees are harvested either singly or in small groups from as little as a quarter of an acre. Regeneration takes place naturally in these small openings. There may be gaps in the forest cover, then, but essentially the canopy is never eliminated. The forest is more diverse in species, more diverse in tree ages and tree sizes. Its proponents argued that uneven-aged management left a better habitat for the red-cockaded woodpecker.

The court's ruling that the Texas national forests were to use uneven-aged management practices was later reversed on appeal. But combined with further Forest Service findings in the late 1980s that the red-cockaded woodpecker was declining in all national forests, *Sierra Club v. Lyng* essentially reshaped subsequent debate on the management of longleaf pine. Now foresters had to ask not only which management practices would stay the decline of the red-cockaded woodpecker, but which ones would actually reverse it? Which forest management practices would increase the number of older trees and thus provide more habitat for nesting red-cockaded woodpeckers?

The Forest Service argued that even-aged management was not just an artificial system hatched in the organizational brain of humans, as its critics maintained, but an imitation of nature. Nature regenerated the longleaf forest by opening the forest with hurricanes, tornadoes, and violent storms that were a constant and terrible presence in the Southeast. Moderate-sized clear-cuts or shelterwood cuttings mimicked the disturbances caused by hurricanes. Indeed, the devastation caused by hurricanes was often astonishing. One traveler making his way through longleaf pine forests from Savannah to Macon in 1828 came across a "tract of country several leagues [a league is about three miles] in width, where every tree was laid prostrate on its side, with its roots torn out of the ground. Their tops were all directed to the south-west, from which circumstance, taken along with various reports in the neighbourhood, I infer they must have been blown down by some furious gust from the north-east." G. Frederick Schwarz, studying the virgin longleaf forests of Louisiana in 1907, found that a devastating hurricane or tornado had nearly eliminated "an old and comparatively dense forest of virgin longleaf pine," carving a destructive path three miles long and a quarter of a mile wide.[9]

The Forest Service believed that such historical evidence proved that the woodpecker had evolved within an uneven-aged mosaic of even-aged stands. As further proof of the compatibility between woodpeckers and clear-cuts, it cited the fact that clear-cutting had been used for decades on the Francis Marion in the years before Hurricane Hugo, on the Apalachicola National Forest in Florida, and on the Vernon Ranger District of the Kisatchie National Forest in Louisiana—three national forests with the largest and most viable red-cockaded woodpecker populations.[10]

Arguments for the uneven-aged structure of longleaf pine were buttressed by the research published in 1988 by William Platt and his colleagues at Louisiana State University. Platt tagged, aged, mapped, and measured almost ten thousand trees within a nearly 100-acre study plot on the Wade Tract Preserve in Georgia. He wanted to study the population dynamics of an old-growth longleaf pine forest—where the old trees were and how they related to younger trees. Platt found that the juvenile and adult longleaf pine weren't spread evenly across the landscape. Rather, seedlings, juvenile trees, and subadults tended to grow together in small but dense clumps segregated by ages. Adult trees, on the other hand, which can reach advanced ages of several hundred years, were often found scattered randomly across the landscape. The researchers hypothesized that the heavy needlefall from older trees fueled hot fires that prevented young seedlings from surviving.

Young trees can grow only where the fuel is sparse or where they can find some protection.[11]

The even-aged nature of the younger clumps did not persist over time but broke down. The older veterans tended to be surrounded by clumps of younger trees, showing that the configuration of the forest changed frequently. The denser groups of young trees experienced heavy mortality, leaving, in time, an opening dominated by an ancient survivor. The older and taller trees were more susceptible to lightning strikes and windthrow, and their deaths provided openings for seedlings to grow.

Platt and his colleagues said that an old-growth longleaf pine forest has an uneven-aged structure, unlike the even-aged structure of a shorter-lived, pioneering pine species like loblolly that grows naturally and quickly in dense, even-aged stands following a catastrophic disturbance such as an intense fire or a hurricane. Longleaf pine on the Wade Tract exhibited the same characteristics shared by other populations of long-lived conifers, Platt wrote. They were of "uneven age, with individuals spatially dispersed in small clusters of trees of even age and size."[12]

Typical openings in a longleaf pine forest are created when numerous trees are knocked down by a tornado, hurricane, or insect infestation. When these trees fall, they make gaps in the forest canopy not only because they knock down other trees, but also because the downed trees become fuel for hot fires. A small opening seeds in from surrounding trees when the logs are consumed, creating teenage ensembles of thickly growing longleaf. The openings caused by the typical blowdown are a few acres or smaller. The larger blowdowns described in the historical record were much rarer.

Yet the dynamics of regenerating a gap of any size in the longleaf pine forest are much more subtle than the simpler scenario in which disturbances cause trees to die and new seedlings to grow in the openings. In a vast opening caused by a hurricane or tornado it's one thing to say that the prostrate forest of several hundred acres will be succeeded by an even-aged patch of young trees and quite another to say exactly how regeneration will occur in an opening this large. The conditions would have to be exactly right. If a bumper seed crop had occurred just before the hurricane and a fire had prepared the seed ground, the seedlings would be ready to take advantage of the light and space suddenly provided. But if the seed was still in the cones, or if a fire had not prepared the seed bed, seeding might not occur and the large opening might persist for many years — possibly even a century or more. The mature, cone-bearing trees on the peripheries of the opening would drop their heavy seeds close to the parent tree. The opening would remain open until the seedlings on the peripheries matured and dropped their own seed, marching inward in this way until the opening was closed. The potential longevity of the longleaf pine tree—approaching five centuries—would enable the forest to accommodate such large, open patches without the forest falling apart.

Indeed, the forest that Platt describes is one of nearly constant motion and random instability, seen on a long-term time scale. If you took a high-altitude aerial photograph of an old-growth longleaf pine forest, you'd see what seems to be a mosaic of even-aged groups of trees within a multiaged forest. Now imagine switching on a camcorder from this vantage point and letting it run for a few centuries—the still picture would become animated with movement and change. Open spaces today might be dense with new growth in 2050. A brushy gang of teenage saplings hanging out in one spot gradually produces a few large, adult trees. Fast-forward the video from the beginning and the forest would look like a squirming, living organism, opening and closing, constantly reshaping itself, constantly reconstituting itself into different patterns. As the centuries pass, the forest shape shifts in response to the laws of its own ecology and to the pressure of local events—windstorm, hurricane, insect attack. Though changing all the time, the forest is always the same.

Ecologists generally believe that the subtle population dynamics described by Platt provide a more "natural" model for longleaf pine management than the hurricane model advocated by Forest Service researchers. A number of scientists connected with the Tall Timbers Research Station challenged the Forest Service's environmental impact statement on the effects of its management plans on the red-cockaded woodpecker. One of them, Todd Engstrom, had studied the red-cockaded woodpecker populations at the Wade Tract and on other large quail plantations in the Thomasville area and agreed that disturbances occurred at different scales in the original forest, resulting in large, even-aged stands, smaller even-aged stands, and a mosaic of uneven-aged patches proceeding from lightning strikes or windthrow. From a management perspective, he wrote, "it makes more sense to mimic [the latter] disturbance regime rather than a disturbance regime based on hurricanes or tornadoes which will, in any event, continue to occur on the landscape at a frequency approximating the past."[13]

And if the Forest Service wanted examples of forests managed successfully according to uneven-aged principles, there were several to choose from.

On a crisp February morning, Leon Neel drove me over to Greenwood Plantation just outside Thomasville, Georgia. For more than a century the 5,200-acre plantation has been one of the jewels of the Thomasville area. Architect Stanford White described the original house, begun in 1835, as "the most perfect example of Greek Revival architecture in America." It was destroyed in a fire in 1993.[14]

Neel was hired by Herbert Stoddard in 1950 when he had just graduated from the University of Georgia. By then, Stoddard was getting on in years and needed an assistant. Neel was a Thomasville native whose family had owned land in the area for several generations. It was a good match, and by 1963 he was a full partner in Stoddard's forest consulting business and continued it after Stoddard's death in 1970. Neel had been a consultant on numerous large properties in the Thomasville area, and many foresters and ecologists respected his forest management style. His management ideas were spread mostly by word of mouth, by means of the field trips he hosted on the quail plantations that he managed and occasions like this one—Neel alone with a reporter proudly showing off his forests.

For the previous owners of Greenwood Plantation who employed Leon Neel for almost fifty years, the chief management goal for these acres was always to produce huntable populations of bobwhite quail. Yet the word "beautiful" was as much a part of the owners' desires as quail or timber.

At Greenwood, we walked into an old-growth section of the plantation called "The Big Woods." I took a snapshot of Neel standing with his arms folded next to a large longleaf. He was very particular about which way he wanted the camera to point, which vista would give me the most pleasure to look at when I returned home. When I look at the photo now, I see an open forest, morning light dappling the golden grass, the trees widely separated, except for a crowd of spindly saplings standing off by themselves. Bushy green seedlings dot the openings. The sun casts a long shadow from the tree he stands near.

"This is a part of a virgin stand here," Neel said. "I call it virgin because the soil is still intact and it hasn't been through cultivation. This is within our commercial area and was logged through this past fall. We took a few trees out. You can see the tops.

"If I could turn all the forest land that I work on into this, I'd be very happy. We're in a grove of older trees—see the flat tops? They're trees of extreme age and extreme character. We've counted rings on trees out here that show that some of these trees might be 300 to 350 years old, and a lot of trees will go 250 to 300. So this is a fine old forest.

"But look at what I'm going to show you, now. Right next to these old trees you'll see some dense stands of young trees and if you look closely you'll see that they are trees of different ages. So it's a multi-age stand. But one thing I would like for you to keep in mind here, the character of long-leaf is such an important part of this forest. The aesthetics are part of the character and the integrity of the forest, not only economic but ecological. The character of this forest changes from hour to hour, day to day, weather system to weather system, day to night, season to season. You can come out here on a rainy day and it's entirely different than it is now. You come here on a moonlit night and it's different, too. Yet it's always beautiful."

Neel was unusual mostly because he maintained that beauty is not an abstract notion for poets and tree huggers, but an integral principle of for-est management — at least the kind of forest management he practiced. His goals on all the hunting plantations he managed were as follows: quail, timber, and aesthetics. "Quail management is a compromise between your multi-purpose goals," he said as we walked down a slope to the bottom, where pitcher plants suddenly announced the wet soils that bathed their roots. "The two most important are timber and quail because timber is money and quail is recreation for the owners. But we also manage for a lot of other things, too. We deliberately manage for red-cockaded woodpeckers. We deliberately manage for orchids and for pitcher plants."

Neel used the word "compromise" many times to characterize his man-agement philosophy. The way he described it, quail management is a con-stant seesaw battle over the proper balance between trees and quail — you can't have maximum numbers of both. "We're not trying to grow the maxi-mum amount of cellulose we can grow on this land," he said, distinguish-ing what he wanted to do here from commercial forests. Some plantation owners believed that the more trees you cut the more quail you had, a posi-tion Neel disagreed with. He fought with them often about it, and some-times he lost. "The bottom line is they really want to cut the timber and they want to justify cutting the timber," he told me. "They want to convert their trees to dollars."

Decisions like these had ecological results for Neel. The more trees that were harvested, the fewer pine needles would be on the ground. He needed the needles to carry the fires, the fires to clear the brush and open the for-est, the open woodland to feed and shelter the wildlife. The less fuel, the cooler the fires and the easier it would be for the oaks to grow up and fire-proof themselves with nonflammable leaf litter.

Despite the ecological thinking that guided his management decisions, Neel had no illusions that the result was anything more or less than manipulation. He planted quail food patches throughout the forest because he wanted artificially high populations of quail. "If you had nothing but woods, with no fields and no openings," he told me, "you wouldn't have quail habitat. Especially at this time of year, the hens have the young chicks out and the open areas have more insects than would open pine woods." Later, he let the patches seed in to secure forest regeneration. Though it looked natural, though it had many old trees, the forest, then, was not simply a natural system. It was man-altered, man-shaped, its shape directed by the landowner's goals.

In Neel's system of harvesting, there were no cutting schedules and no rotations. The forest was not divided into compartments for easy organization. There were no year classes. Neel knew his forest and marked individual trees for harvest. He followed Stoddard in using a form of single-tree selection. Neel frequently said that his marking philosophy was aimed more at leaving trees than harvesting them. He removed one because it was diseased or defective in some way. He took out another because its absence would allow a better tree to put on more growth. This old snag was left because of the shelter it offered woodpeckers, owls, and other cavity dwellers. This mature veteran, which could be cut, was left because it was a prolific seed producer. The trees were not dragged to a spot where they were cut, possibly destroying regeneration in the process. They were limbed and cut into logs on the spot.

The result was a carefully managed, profitable longleaf pine forest. At Greenwood, between 1945 and 1975, Neel and Stoddard cut a million board feet per year from the plantation while doubling the volume of timber on the ground, a high yield for a management system with multiple objectives.

Despite this, Neel's single-tree selection system has never impressed the Forest Service. Bob Hooper, a retired wildlife biologist formerly stationed at South Carolina's Francis Marion National Forest, was particularly scornful of single-tree selection, while admiring the artistic forestry of Leon Neel. "I like Leon tremendously," he told me. "I think he's a good forester. But when I hear people advocate for public lands what Leon is doing for private lands, I kind of go ballistic because I don't think they have a clue what the hell they are talking about. It would be inappropriate to manage public lands the way he does it in Thomasville."

Hooper's problems with Neel's single-tree selection method were based mainly on its haphazard effects on forest regeneration. Quail plantations provide a poor model for forest management, he argued, no matter how well they're managed, because hunters prefer open hunting courses so they can see the birds. There will be far fewer trees on a quail plantation than on a forest for which timber is still a powerful objective. "If you're lucky you get some regeneration in a hole left by a tree. But if you have public land with multiple goals, like it or not one of these goals will be growing trees for commercial purposes. And in that case, you don't want to do anything close to single-tree selection."

Hooper also believed that to "manufacture" his picturesque forests, Neel was essentially cheating by eliminating the young growth that looked rough and unkempt, thus endangering the future of the forest. To keep his wood-lands open and regal looking, he burned them annually and thus killed too many young trees. "He's destroying his regeneration because quail hunters don't want to shoot through a lot of bushy young longleaf pine growth," Hooper said.

This was a frequent criticism of Neel's single-tree selection system and one that stung him a bit. "Most foresters come up to Greenwood and say, where's your reproduction," he told me. "I say, there are two things to con-sider: first, we don't really need it because of all the timber we've got on the ground. But the second thing is, it's out there. All you got to do is look and see it. But you know, reproduction to them is rows and rows of even-aged young trees over hundreds of acres."

He accepted the compromises involved in managing for quail and timber and woodpeckers and pitcher plants and beauty because he knew that his management did not jeopardize the existence of the ecosystem — the known and unknown ecological relationships still functioned, the natural processes such as fire still worked, the forest was still reproducing. "Our objective is to keep this forest going and this land going, going forever if possible, not just for the benefit of a short-term thing," he said.

In July 2002 The Nature Conservancy was awarded management of the 5,200-acre plantation on the death of Greenwood's owner, Mrs. John Hay Whitney. Full ownership was expected to be transferred to the organiza-tion within a year. The Conservancy credited the innovative management of Herbert Stoddard and Leon Neel for the condition of the old-growth longleaf forests on the property, which contained significant populations of red-cockaded woodpeckers, gopher tortoises, and other rare species. Neel continued as the consulting forester on the property.

In the timber town of Brewton, in southern Alabama, I visited another forest where successful uneven-aged management had been demonstrated for many years. I spent a weekend with forester Bob Farrar touring the Forest Service's Escambia Experimental Forest. Farrar was a retired forest researcher with the Forest Service, a self-proclaimed forestry heretic, a good ol' boy from Mississippi with a Ph.D. in forestry and a machine-gun delivery. Just to confuse the image even more, he was driving a red Nissan 300 SX when I met him, somehow folding his lanky frame into the car like a carpenter's ruler.

One other thing: Farrar has written some of the most formative papers on longleaf pine management of anyone working in the field. He was also an advocate of uneven-aged management for longleaf and a ferocious critic of the Forest Service, which he referred to as "Smokey Bear."[15]

Farrar worked for thirty five years in southern pine silviculture, most of that time at Escambia, and, along with Tom Croker and Bill Boyer, helped lay the foundations of successful longleaf pine silviculture. Later in his career, he saw how uneven-aged management using group selection could work with longleaf, and since then he has promoted its potential for landowners large and small.

"Smokey Bear, when he first got pressured about all the clearcutting he was doing, imitating industry, he fell into one of its party lines," Farrar told me as we toured some of Escambia's even-aged and uneven-aged forest stands. "Smokey Bear said, 'Southern pines are even-aged species and cannot be managed any other way.' They made that bald-faced statement. That shows ignorance. You can manage these species any number of ways. Some are good, some are better than others."

Farrar's method of uneven-aged management is often lumped together with Leon Neel's single-tree selection method, but the two are very different. Farrar worked with small groups of trees, a harvest method called "group selection." He was a connoisseur of patches and holes. He looked for holes where regeneration had already come in. He might enlarge a hole, removing several trees to enable the young seedlings to grow faster. Elsewhere, he'd remove a patch of big trees where younger trees had gotten started. Farrar didn't kneel at the altar of old growth. He identified a maximum diameter at breast height, which guided his cut. Not everything above this diameter was cut; the maximum was flexible enough to accommodate the needs of woodpeckers and other species dependent on old trees, but he found no reason to keep trees growing for two hundred or three hundred years.

In Neel's forests, the occasional big, old tree or groups of old trees provided an aesthetic that Neel allowed for. A forest that looked beautiful and mysterious and old was a goal almost as important as how many board feet could be cut or how many quail could be raised. Neel's method was also much less quantifiable than Farrar's. It's a man walking in the woods, sizing up this tree against that one, looking at vistas and trying to figure out how to keep them open, thinking into the future, weighing the need for quail against the need for timber, and trying to balance them both against the need for wiregrass, pitcher plants, red-cockaded woodpeckers, gopher tortoises, Bachman's sparrows, and other distinguished members of the longleaf ecosystem. "A certain art is required in making this judgment," he said in an understatement.

Farrar disparaged it as "crystal ball gazing." When I asked him the main difference between his method of marking trees for harvest and Neel's, Farrar replied that his method was teachable, Neel's was not. "I can take you out and in a half day you'll know as much as I do. With Leon it's a black box, and he ain't going to open the lid on that box to show you what's going on inside."

At Escambia, we looked at a variety of plots, some even-aged, some uneven-aged. As we eased the car off the road under a big pine, Farrar said: "One of the ways you can tell the northerners from the southerners on a tour is that we'll park our cars in the shade. The Yanks park the sons of bitches in the sun!"

We looked at an even-aged plot that had been managed using shelterwood cuts. When you walk into a shelterwood stand, you often won't find any seedlings on the ground. A good shelterwood cut will leave a stand remarkably clean, the seed trees spaced out across the site. I remarked to Farrar that I couldn't see much regeneration here. He said it wasn't necessary yet. In even-aged stands such as this shelterwood, you don't capture the seedlings until you're ready to harvest. Five years before you plan on harvesting the stand, you do an initial cut to reduce the number of trees on the stand. That's when you wait for the seeds to drop.

"Even-aged stands have a definite tenure on earth," he said. "They are born, they grow, and they die. But before you can take the trees off, a lot of critical things have to happen. You've got to get the competing hardwoods down to zero, you need to have a decent seed crop, you need to get the overstory off as quickly as possible so you can get the seedlings into some height growth and away from the next wave of hardwoods that are damn sure going to come back like death and taxes, okay?"

Down the road we looked at an uneven-aged plot that was a chaos of little trees and saplings and big trees. "You can see the little clumps and patches here," Farrar said. "There's a group here, some pulpwood size out there, and an even-aged patch of big timber right here and another size class over there. Look at those big rascals over yonder!

"In an uneven-aged stand like this there's no beginning or end point. Regeneration and cutting are going on all the time. Each patch is going to regenerate, grow to sapling size, move on to pulpwood size and while that's going on other patches are moving along as well."

Did he keep track of all the age classes? That's one of the frequent questions that forest managers had asked him on past tours at Escambia. Managers of even-aged forests have to know exactly what is happening on every stand so they can schedule a harvest cut or a burn. Scheduling is the whole point of even-aged forest management. Organizing or "regulating" a forest into different age classes helps traditional foresters, whereas a mixture of age classes on a single plot often confuses them.

"Relax," Farrar would tell them. "All you do is inventory your stand. If it's grown enough you cut it, and if it hasn't you don't. See, this is ideal for private landowners who don't have to be on a tight schedule."

For Farrar, the need for scheduling was the strength of even-aged forestry but also its weakness. "Time is on our side in uneven-aged management," he said as we wandered through the stand. "If we keep burning, if we keep cutting, the seed will fall, some of it will catch, some will burn up, the birds and rodents will eat up a whole bunch. The majority of the seed is lost but we get enough to keep the patches and the groups coming up. We substitute cutting for natural mortality."

He pointed to a few tall trees with young seedlings coming in underneath. "See, that's a clump of trees that we can take out and get a whole new bunch of trees started. They're all out there waiting to take off. There's no hard and fast rules about how big the patches have to be. That's what destroys the bean counters."

There were disadvantages to his group selection method, he told me. It cost more and it required a level of skill that was often beyond what the Forest Service could provide on the ground. "They don't have to be geniuses, or Ph.D.s, but they have to know what the hell they're doing," he said. "They've got to be trained. You don't put a truck driver in a damn airplane and expect him to take off. But he could learn to fly that son of a bitch!" Because of the Forest Service's huge acreages, he admitted that even-aged management using the shelterwood method, which was pioneered here at Escambia, made sense. But using the shelterwood method required a technical proficiency that Farrar believed was missing among foresters on national forests today. Even shelterwood had been a late entrant into the Forest Service's management manuals, despite the research that Tom Croker and Bill Boyer performed at Escambia. "It was twenty years before [the Forest Service] took the seed tree method of regeneration out of their manual and got the shelterwood method in there," Farrar said. "Twenty damn years after we knew that the seed tree method didn't work they were still doing seed tree in longleaf and losing it to hardwoods!"

The advantages of using uneven-aged management included a bigger payoff for the red-cockaded woodpeckers—and landowners—because it produced larger trees, exactly the ones that the red-cockaded woodpecker needed; it also earned the landowner the most money. It was easier because it required less work. "Uneven-aged management is not a pulpwood producer," Farrar asserted. "A lot of uneven-aged forests are never cutting a piece of pulpwood out of it; they don't worry about it until it gets up to sawtimber size. Managers don't care about how old it is. It's a hidden inventory that they don't worry about. They just check every once in a while to make sure it's still there. Whereas when these even-aged stands get to be about twenty-five years old, you've got to go in there every five to ten years and be doing something to each one of them."

Though enthusiastic about the prospects of uneven-aged management and group selection for growing longleaf, he was pessimistic about the long-term prospects for recovering the red-cockaded woodpecker. "Let's face it, folks, the red-cockaded woodpecker is gone," he declared. "It's just a matter of time. We're going to kill him and a whole lot more before it's over with. There's too many of us on too little earth."

Which is not an opinion shared by the U.S. Fish and Wildlife Service in its second revision of the recovery plan for the endangered woodpecker signed in early 2003. The plan highly recommends uneven-aged management of longleaf to produce habitat for the bird, either with the single-tree selection method of Leon Neel or various group selection methods, including Bob Farrar's. It does not recommend even-aged management methods —clear-cutting, seed-tree cutting, and conventional shelterwood cutting. Clear-cutting is approved only to convert off-site pines like slash and loblolly pine to longleaf. The recovery plan blesses a modified form of shelterwood whereby the seed trees are never harvested and rotations are extended to 120 years to provide the old trees that the bird needs.[16]

Critics have been skeptical of the modified shelterwood system because leaving mature trees permanently would make them susceptible to wind storms and reduce growth in the young trees by 50 percent. Farrar agreed. An irregular shelterwood won't work, he said, because it's precisely those big, older trees sticking up in a patch of young trees that run the highest risk of being killed by lightning or tossed over by strong winds.[17]

"If they want to make sure that they can keep trees for the bird, the best way to do it is use this uneven-aged system," Farrar told me. "You need to keep growing those big trees right where he is. You leave them and write them off. Your forestry is not as efficient but you always maintain suitable trees for the woodpecker. What suffers is wood production. That's a concession to the bird."

The debate over forest management methods in longleaf pine forests suggests again how young forest management of any kind really is and how it continues to evolve. The Stoddard-Neel single-tree selection method has been practiced for about sixty years, yet that's not a long time in the life of a forest. Neither is Farrar's group selection method, which has been tested at Escambia only for the last twenty-five years or so. The goal of even-aged and uneven-aged management is a sustainable forest in which a percentage of new growth can be removed periodically. The goal, furthermore, is to maintain this process perpetually. But no one can be totally confident of the success of any management system, and that includes the even-aged practices on intensively managed pine plantations. Will highly manipulated tree farms continue to produce timber in the quantities desired? Or will they succumb to the increasing need for more chemical fertilizers, insecticides, and herbicides?

On the other hand, there is great hopefulness in many quarters over the positive effects of red-cockaded woodpecker protection on the future of longleaf, especially on public lands. "I believe that on national forest lands in the Southeast, we have turned the corner in the loss of longleaf," said Forest Service forester Ron Escano at the 1991 Tall Timbers fire ecology conference. "We will see a significant increase in the amount of acres of longleaf in the next decades." As the debate continues about the merits of even-aged or uneven-aged management, longleaf pine today is being managed more successfully on national forest lands than anywhere else.[18]

Leon Neel agreed, with some sadness. Though the large privately owned hunt clubs of southern Georgia and northern Florida at one time practiced some of the best longleaf pine management, a new generation of owners lacks the land ethic of their parents and grandparents, and their disposable wealth, as well. "At one time I felt that the future of the longleaf pine ecosystem rested in the hands of the private landowners," he told me. "At the time we were dealing with Forest Service people who were the worst land managers in the world. But I've changed my mind. I think the Fish and Wildlife Service and the Forest Service are the ones that will perpetuate blocks of forest land large enough to salvage some of the ecosystem."

FOURTEEN

Restoring an Ecosystem

Humanity must begin to view itself as part of nature rather than the master of nature. It must reject the belief that nature is ours to use and control.
—T. R. Stanley (1995)

At century's end, the restoration of longleaf pine on public and private lands had become a major forestry goal in the South, although there was confusion over exactly what "restoration" meant. Some described the goal as restoring the "original ecosystem" or "presettlement conditions." Yet this phrasing rang warning bells for many foresters and ecologists. Which "original" ecosystem was to be restored? The eighteenth-century forest before the loggers and turpentiners had despoiled it? The fifteenth-century forest that existed before the Europeans arrived? Or the forest of 12,000–15,000 years ago that preceded the human entrance into North America? Each forest was different. Hadn't the Native Americans changed the forest they found just as the Europeans changed the forest that the Native Americans had shaped? And weren't landscapes dynamic entities, changing naturally over time as a result of short-term natural disturbances, long-term climate changes, and even longer evolutionary time scales. As ecologist Reed F. Noss phrased it, "The presettlement forest in a particular landscape was one frame in a very long movie, played on a projector with no reverse switch."[1]

In such a dynamic world, even individual species might be seen as having no right to permanent tenure, and efforts to protect endangered species from extinction could be judged as sentimental. From this point of view, landscapes, ecosystems, species—all are mere flotsam on the vast ocean of time, whirled away in a matter of a geologic moment.

If the term "original ecosystem" was dubious and "presettlement conditions" were just another stage in the long cavalcade of change, then what were forest managers going to restore?

North Carolina ecologist Cecil Frost grappled with the issue and defended the concept of "original vegetation." Our climate has been relatively stable for about 8,000 years, he wrote, and most contemporary plant communities have been in place for 6,000 years. "Given that these natural communities existed for all of human recorded history, it seems reasonable that these are the communities that we would want to perpetuate on natural areas." Noss also tried to rehabilitate the concept of restoration. Although it wasn't possible to return to some "pristine natural condition," he argued, you could attempt to reverse the "trajectories of impoverishment" that have degraded contemporary ecosystems. That in itself would be restoration. "The restored, sustainable forest will not be the same as the pre-European settlement forest, but it will be closer to it in some fundamental ways than the present, exploited forest."[2]

This was a more pragmatic perspective, and Forest Service researcher Joan Walker adopted it in her own views on the subject. As a member of the Forest Service's Longleaf Pine Ecosystem Restoration Team, she put landscapes on a continuum between the "highly modified (e.g., parking lot)" and the "very natural." She defined "restoration" as any action that moved a site closer to the natural end of the continuum. "This model avoids the problem of precisely defining natural or specifying a point in time as the restoration objective."[3]

A similar pragmatism was expressed by the Longleaf Alliance, a nonprofit organization founded in 1995 as an information clearinghouse for longleaf pine growers. Codirectors Rhett Johnson and Dean Gjerstad asked, "What do we mean when we talk of restoring the longleaf forest? Do we mean getting longleaf back on the land or do we mean restoring longleaf along with the rich plant and wildlife community we commonly associate with fire maintained longleaf forests?" Given the diverse conditions of longleaf lands in the South, ranging from degraded to intact, Johnson and Gjerstad decided that the restoration could embrace the simpler goal of "getting longleaf back on the land" and the more complicated goal of getting it back on the land "along with the rich plant and wildlife community we commonly associate with fire maintained longleaf forests. . . . Longleaf is better than cotton; longleaf with wiregrass and native legumes is better than longleaf alone; longleaf with wiregrass and native legumes and quail is better than longleaf and native ground cover alone; longleaf with native groundcover, quail and gopher tortoises is even better, etc."[4]

This concept of a restoration continuum made sense of the diverse and uncoordinated efforts to restore longleaf pine that I found on my travels through the Southeast. A little 100-acre project to plant longleaf pine for pine straw could appear on the same restoration continuum as Leon Neel's artistic management of the Greenwood Plantation's ancient longleaf. And somewhere on the same continuum lay the more ambitious goals of *ecosystem* restoration that I was witnessing on many national forests and even on some private lands.

Indeed, the changing emphasis in the forestry world from timber production to ecosystem management was an issue that was beginning to stir ecologists and foresters at the end of the twentieth century. For most of the century, forests had been described as a resource that furnished "goods and services" and forestry was the science of producing those services. "In particular, it is the art of handling the forest so that it will render whatever service is required of it without being impoverished or destroyed," wrote Gifford Pinchot in 1914. Depending on their objectives, forest managers could increase good hunting opportunities for bobwhite quail, as Leon Neel was doing in Georgia, or they could manage the forest to produce timber, which is what the U.S. Forest Service defined as its main purpose for many years. Managers might even choose to do nothing, a choice that many landowners in the South make by default.[5]

But was it also possible for managers to enhance all of the forest components, including quail, timber, wiregrass, and red-cockaded woodpeckers? In this scenario, the manager's goal would be to keep the ecosystem functioning, not to manipulate it to produce desired products. As wildlife biologist Jay Carter told me: "People try to treat [longleaf pine forests] species by species—they try to go into this little tract of land and manage it for the red-cockaded woodpecker and that one for some endangered plant and that little patch for quail. They cut a little timber. If you manage the whole system as a natural system, the other things will take care of themselves."

That was—and still is—the promise of ecosystem management, the new forest management paradigm that has guided the U.S. Forest Service since 1992. At the core of this idea is the belief that the best forest management maintains the full diversity of an ecosystem's organisms—trees, grasses, herbs, fungi, wildlife—as well as their multiple and complex interconnections. As the Ecological Society of America defined it, "Ecosystem Management does not focus primarily on the 'deliverables' but rather on the sustainability of ecosystem structures and processes necessary to deliver goods and services." From this perspective, the history of forest management in America can be seen as a growing acknowledgment of the complexity of natural systems and the sheer difficulty of the management enterprise.[6]

As scientists acknowledged the complexity of the systems they were trying to manage, some profound disagreements emerged over the hubris in-volved in the management of natural resources. W. G. Wahlenberg had wor-ried over the fallibility of forest managers when he admitted in 1946 that the mismanagement of longleaf pine forests had been the norm rather than the exception. Paul Ehrenfeld's influential book, *The Arrogance of Humanism*, published in 1978, advanced the sense of a widening gap between humans' ambitions to control nature and their abilities. "We must come to terms with our irrational faith in our own limitless power and with . . . the widespread failure . . . of our inventions and processes, especially those that aspire to en-vironmental control," he wrote. An increasingly skeptical critique focused on the limitations of human knowledge in the face of the complexity of natural systems. It was common in the 1990s to run across such statements as, "It seems doubtful that a profound shift in ethical consciousness can be achieved unless humans develop a far greater sense of humility, respect, and even awe toward nature" and "Humanity must begin to view itself as part of nature rather than the master of nature. It must reject the belief that nature is ours to use and control."[7]

Given the historical background of longleaf pine forest management, it seemed to some that the new rubric of ecosystem management verged on the presumptuous. Could we guide and maintain all those mysterious re-lationships and processes that comprised a functioning ecosystem? Did we really know enough to *manage* an ecosystem?

———————

I was at the biggest sandpile in the Florida Panhandle. Eglin Air Force Base is a brute of a military installation, nearly 464,000 acres broad and 724 square miles in extent, the largest military base in the country. If you traced an oval within the area bounded by the towns of DeFuniak Springs and Freeport on the east, Pensacola on the west, and the trending lines of Interstate 10 and Choctawhatchee Bay on the north and south respectively, you'd have the approximate shape of Eglin. The base includes twenty miles of beachfront property along barrier islands stretching over three counties.

Eglin is the site of the former Choctawhatchee National Forest, where for four decades foresters cut their teeth on the complexities of longleaf pine management, and where they eventually acknowledged their failures. Today it's a place where U.S. Air Force hotshots practice their bombing runs and test new weapons systems.

Nearly 80 percent of the base is upland sandhills, and nearly 100 percent of that was in longleaf pine a century ago. Today, the bulk of Eglin is still forested, but most of it is a substitute forest of sand pine, slash pine, and turkey oak that grew up partly by intent and partly by accident.

Yet things are changing at the former Choctawhatchee National Forest. The Eglin Natural Resources staff has been imbued with ecosystem management ideas, and their work on the base has won them admiration and national recognition. I visited Eglin to see what they were doing. Forester Scott Hassell was my tour guide for part of the morning. In little more than an hour we made quick windshield surveys of nearly a dozen sites where sand pine and slash pine had grown up. At some of these sites, signs announced "Ecological Restoration in Progress."

The restoration in progress consisted of removing between 60,000 and 90,000 acres of the slash and loblolly plantations at Eglin, perhaps 2,000 acres per year. "We're converting these plantations of off-site pines to longleaf pine," Hassell told me at one small site. "Here we're hand-planting longleaf seedlings on ten to fifteen acres without traditional site preparation and leaving the overstory, things that from a commercial point of view were not recommended. To tell you the truth, it was an unorthodox procedure and I was skeptical about it."

Hassell, like many foresters, was educated at a time when the goal of forest management was chiefly to produce timber and other commodities. To do that, foresters clear-cut their forests and planted new plantations of young loblollies or slash pines, eliminating competing vegetation with chemical herbicides. The new prescriptions at Eglin made him a little nervous at first.

"You know, all of the research that has been done on longleaf pine has been done from a commercial point of view," he said. "We had to redefine our views on what success was. Is it 95 percent seedling survival? It might be 25 percent or 50 percent."

He pointed at some of the longleaf seedlings struggling out of the sandy soils. "They're only four years old and they're growing slowly, but they're starting to grow."

As we drove slowly from site to site, Hassell explained that the sand pine being clear-cut was originally a tree of the coast, although it grew in some of the scrub areas on the base as well. One variety of sand pine has a serotinous cone, releasing its seeds in intense fires. At Eglin the variety of sand pine prolifically dropped seed every two years, whereas longleaf pine on its sandy soils produced a seed crop every ten years or longer. Because sand pines are shallow-rooted, fire is their great enemy, and the frequent fires in the longleaf pine forests had originally confined the sand pine to the scrub. But when fires were suppressed earlier in the century, the sand pine's light seeds were carried far and wide by wind and the pines began to encroach onto the cutover longleaf stands, shading out the ground cover grasses and plants and diminishing the diversity of the forest. Although there was hardly any sand pine on the site when the century began, in the ensuing decades it became a larger and larger presence, expanding to 60,000 acres by the early 1990s and covering nearly 20 percent of the former longleaf pine sandhills.

The small number of slash pines that had originally hugged the wetter areas of the base had also expanded. In the years after the War Department took over Choctawhatchee in 1940, Eglin's timber managers succumbed to the lure of the miracle tree and in the next thirty years they planted 30,000 acres with it. In some few areas where the soils were wetter, slash pine grew well, but in the predominantly deep sandy soils of the base, the slash pine had dashed from the blocks like a sprinter and then had given out yards short of the finish line. At Eglin and in many sandy places all over the South, thirty-year-old slash pines stood stunted in dog-hair plantations, showing little or no growth over the last decade.

Eglin's forest managers were removing them, taking what they hoped would be the first sure steps toward restoring what grows best on these deep sandy soils—longleaf pine. It could take as long as twenty years to cut out the slash and sand pines, Scott said, although they had to make sure that their removals wouldn't hurt the endangered red-cockaded woodpecker, which had built cavities in some sand pines on the base.

Managers were haunted by the irony that most of their activities so far had amounted to little more than attempts to undo the results of previous mismanagement. "We're trying to find out what we can do as quickly as we can, without adversely affecting the environment or the military mission," Hassell said. "We're doing research on our methods here and what they're doing to the ecosystem, and we're adapting as we go."

As big as Eglin is, it is only a single member of an even larger ecological unit, an organization called the Gulf Coastal Plain Ecosystem Partnership. This group consists of eight public and private landowners of contiguous longleaf pine located in three counties in the northern Florida Panhandle and southern Alabama. Along with Eglin Air Force Base are Blackwater River State Forest, Blackwater River State Park, Conecuh National Forest, International Paper, Northwest Florida Aquatic Preserves, Northwest Florida Water Management District, and The Nature Conservancy. Partnership members manage a total of 927,295 acres and represent the largest contiguous block of longleaf in the South. With 75 percent of the remaining longleaf in stands of less than 100 acres, the giant partnership is creating mind-boggling educational and research opportunities. Members have successfully joined in several cooperative management ventures, including prescribed burning over their joint landscapes, monitoring of red-cockaded woodpeckers, and public education.

Joint forest management of this size and scope is unprecedented in a southern landscape increasingly fragmented by roads, development, and private ownership. Management on a larger landscape scale offers better connectivity and increased genetic diversity among wildlife populations, more protection to endangered species, and more scope for natural processes such as fire and storms. The 200-acre Wade Tract Preserve, as valuable a model for managing old growth as it is, is so small that the naturally changing mosaic of forest openings and closings that characterize the ecosystem may be slowed or even ended at some time in the future.

As we drove along a sandy ridge road off Eglin's main road, I was suddenly aware that we had been passing dozens of flat-topped longleaf pine trees, the sure sign of an old-growth forest. As I looked, the dozens turned into hundreds, perhaps thousands. We were driving through one of the largest assemblages of old longleaf pines that I had ever seen, but the pine forest had been difficult to see because it was wrapped up by mean-looking thickets of large-trunked turkey oak. The tops of the stunted longleaf barely rose above the turkey oaks.

"Good Lord!" I exclaimed. "Look at this!"

I was full of questions, but Hassell had stopped the truck and suggested that I join Carl Petrick, who had been trailing us for some time. Petrick was the wildlife biologist at Eglin, a stocky man with an infectious way of talking about Eglin and its new management ideas. He had spent some time that morning briefing me on the management history of the base and showing me maps, and I had grown to like his optimistic views on longleaf pine and its future at Eglin. Hassell had other matters to attend to, so I jumped into Petrick's truck and peppered him with my questions.

"Tell me about this place," I said. "How old are these trees? How many acres is this?"

"We call this Range 62," he explained. "It's a long sandy ridge, about 300 to 400 acres in extent, and it's all old-growth. We've never cored these trees to find out exactly how old they are, but I can guarantee you that many of them are in excess of 350 years old. When we first wrote our management plan, we had no idea of the density of the trees here. I mean it was a sea of turkey oak mid-story, and we couldn't see twenty yards into that stuff. We're just getting a sense of the age of some of these trees. We have so many researchers on contract now, and they're coming back all the time telling us, 'I cored a tree that was nine inches in diameter and it was 350 years old!'"

Researchers are now saying that Eglin has more old-growth longleaf pine than any other area in the South. The Patterson Tract, another old-growth sandhills site I visited later that day, contains 1,000 acres of ancient longleaf pine trees, many of them over 400 years old and one of them, the oldest longleaf ever aged, over 500 years old. More than 10,000 acres of old-growth longleaf have been identified on the sprawling base, and the count could go even higher as research continues.

The oaks were still in control on this site, however, because years of fire suppression had allowed them to gain a stranglehold on the longleaf. In some cases the trees were six inches in diameter. Restoration activities here consisted of removing many of the oaks with chain saws and introducing growing-season fires to knock the others back. Fallen hardwoods littered the forest now, and some of them blocked the road, forcing us to lug the heavy trunks of century-old oak trees away.

"I'll tell you what, we had to fight some serious battles to get growing-season fires started," Petrick said with a laugh. "The first time you have a growing-season fire, the needles are burned and they turn brown and cover the ground. You hear comments like, 'What's the objective of this burn, to kill every living thing?'

"But when you see the plants react the way they do to growing season fire — the plants that won't flower and won't set seed unless they're burned during the growing season — when you see how those forests look and you see healthy populations of rare plants and rare animals and an overstory of older longleaf pine and an understory of younger longleaf and a lush ground cover, I guess you'd have to be an idiot not to put it together."

As we gazed over the ancient pine forest that was only beginning to emerge from its oak prison, I began to see it as a peculiarly apt metaphor for longleaf pine's predicament at the beginning of the twenty-first century. At Eglin and at many other places across the South, progressive managers were finally beginning to liberate longleaf pine forests from the effects of destructive turpentining, cut-and-run logging, and forest mismanagement. The beleaguered forests were also figuratively escaping from a reputation that almost had doomed them to extinction. For a moment I thought of *The Prisoners*, Michelangelo's unfinished sculpture that depicts tormented men struggling out of their marble prison.

I drove from Baton Rouge east to Camp Whispering Pines, a 600-acre property owned by the Girl Scout Council of Southeast Louisiana. My route led me through country that failed to hint of Louisiana's remarkable longleaf pine heritage. The railroad loggers had not only obliterated longleaf from the landscape but also seemed to have left a flattened and undistinguished countryside in their wake, marked by the same listless small farms, auto repair shops, chain motels, and fast-food restaurants that you could find anywhere in the South.

An abrupt change occurred as I approached the main entrance to the camp. Suddenly I was aware of open pineland with tall longleaf pine trees and flourishing grasses. It was so unexpected that I drove past the gate just to gawk. Later, when I described my pleasure at the scene to Jean Fahr, assistant executive director of the Girl Scout Council, she laughed. "A lot of people tell me that they instinctively slow down when they drive past the camp," she said. "Some of them even tell me they remember when it used to be like this all over."

I had come to Camp Whispering Pines to witness the results of an ecosystem restoration project that the camp leaders, with the enthusiastic participation of the Girl Scouts themselves, had been engaged in since the late 1980s. Earlier in the decade, the camp hired a consulting forester to develop a plan to manage the property's valuable forest of sixty- to seventy-year-old pines with an eye to selling the timber. Remarkably, when the forester actually looked at the camp's forest, he pointed out that such a plan was simply inappropriate on the biologically rich site and that a different approach was needed. Jean Fahr and other camp officials met with the forester and representatives from The Nature Conservancy, and out of the meeting came a proposal for a new forest plan, one that "considered the total ecosystem, including plants and wildlife as well as timber." Latimore Smith, a biologist with the Louisiana Department of Wildlife and Fisheries, enlisted longleaf pine researcher Bill Platt of Louisiana State University to support the development of the restoration plan for the Girl Scouts, based on Platt's well-documented work on old-growth longleaf pine at the Wade Tract.

I called Platt to ask him to show me around Camp Whispering Pines and describe the progress he was making on the restoration project there. Happily he agreed to meet me at the camp.

On the morning of the tour, a heavy gray wall of rain intermittently lashed the blackened columns of pine. It was not a promising day, yet Platt was there with a platoon of graduate students and a mayhem of dogs. In his mid-fifties with a shock of white hair that fell over his forehead, he wore a ripped Nature Conservancy sweatshirt, blue jeans, and white rubber boots that came to midcalf. When it rained, which it often did that day, he opened a large golf umbrella and let it lean against his shoulder as we ambled through the sodden woodland.

"We are making spectacular progress here," he acknowledged. "It has to do with a number of factors. This longleaf community falls into a unique and globally rare longleaf type: fine soils that are mixed with and capped by loess deposits. Thus, it's a very fertile site. It creates a lot of fuel and so the fires burn hot." He explained that in forests growing atop sandy soils, you have to wait a few years in between burns to allow the vegetation to build up again to carry a fire. At Camp Whispering Pines, there was enough fuel for fires to burn just about every year.

As our sprawling entourage of people and dogs ambled through the forest, Platt talked about the history of the area and his own management ideas. When loggers had removed all of the longleaf pine and other merchantable timber from the site, he said, some second-growth longleaf grew back among the stumps along with considerable amounts of loblolly pine. Additional loblolly was established in plantations on sites once occupied by longleaf, and at some point cattle grazed the growing forest. Eventually, under decades of fire suppression, thickets of sweet gums and water oaks grew up, accompanied by exotic privet bushes. When Platt first stepped foot onto Camp Whispering Pines in 1992, the vegetation was so thick that the girls often were unable to see a neighboring tent twenty feet away. The second-growth forest of loblollies and longleaf pines, along with some naturally occurring shortleaf pines, was even-aged, all of the trees about sixty to seventy years old, so densely stocked that there was relatively little reproduction in the areas where longleaf had managed to grab a foothold.

The restoration challenge was clear enough: restore the forest to "an open, scenic, uneven-aged longleaf ecosystem and other ecosystems similar in appearance and composition to those historically present on the site," as the management plan stated. The goals included the maintenance and promotion of naturally occurring wildlife populations and plant cover and the use of the property for educational, recreational, and research purposes. To do this, frequent fire had to be introduced to rid the camp of the privet and other invasive hardwoods. The off-site pines—the loblollies—had to be removed and longleaf seedlings planted.

We looked at a solid loblolly forest with a small clearing visible from the road. The clearing had been made by loggers when they removed a small patch of loblollies. In the patch the girls had planted longleaf seedlings. Platt could have directed the entire area to be clear-cut and replanted in longleaf, but that would have been aesthetically unacceptable to the camp. "Rather than clearing it all at once, we want to open up these loblolly forests in bits and pieces," Platt explained. And there was another reason for the gradual approach. Since Platt's aim is to produce an uneven-aged forest, his openings will occur regularly over many years, enabling the periodic harvests to produce a continual stream of income for the camp. In the future, small patches of uneven-aged longleaf will be growing where the loblolly grows today.

We moved to another area below the dining hall where muddy ruts marked recent logging activity about three months before. To minimize messy logging scenes like this, Platt said he would like to experiment with a small feller-buncher on low-pressure wheels that could fell and cut an individual tree into lengths, followed by an equally maneuverable truck that would load the cut logs. It's a cleaner method of logging that might not harm the ground cover as much as regular logging. Platt quickly swept his hand over the scene. "When I first came here," he said, "I thought this was all loblolly—that's how high the hardwoods were. Our goal here is to open this up, thin all the loblollies and a lot of the hardwoods, burn it and hopefully the ground cover will come back."

If there was anything that distinguished the restoration efforts at Camp Whispering Pines from Mickey Webb's work in Mississippi or Leon Neel's in Georgia, it was Platt's concern for the ground cover. It might even be said that Platt was even more concerned with the ground cover than with the longleaf pine. This emphasis on the entire suite of plants distinguishes ecosystem management from management philosophies that tweak forests to produce favored products, whether they are bobwhite quail, red-cockaded woodpeckers, or timber.

He surprised me from time to time that day by pointing at what seemed to be a beautiful stand of solid longleaf and saying that it was too dense: there were too many trees. I wondered about that because many turn-of-the-century photographs I had seen showed what seemed to be extremely close-growing stands of longleaf, and I remembered as well the advertisements for Calcasieu lumber that yielded 30,000 cubic feet to the acre. It must have come from extremely dense forests.

"There's no doubt that on some of these fertile upland soils the trees got pretty dense," Platt told me. "I have no problem with patches of dense trees, but I don't want the whole thing to be dense. I'm attempting to restore conditions that will favor the ground cover and maintain the biodiversity." His research at the camp had revealed that the biodiversity — the number of plants and the variety of plant species — was richest in the open spaces and poorest under the trees. He said that the biodiversity in the open spaces at the camp rivaled that in other pine savannas along the Gulf Coast, where as many as 130 plant species had been counted in some areas as small as 100 meters square and as many as 45 species in a single square meter.

"That's the reason I'd rather have it somewhat open," he said. "Openings like these would have been natural, created by hurricanes, fires, or beetles. It's a dynamic landscape and so we're re-creating openings and hoping that many of the species in the ground cover that would have been here actually are here and will respond. I want to thin it so we can get some regeneration in here and head toward a more uneven-aged stand. That's my goal."

Along with Jean Fahr and the camp ranger, Larry Ehrlich, Platt frequently monitored areas that had been logged or burned, evaluating the results and making adjustments. For example, Platt was trying to understand the optimum size of the openings that he was creating. If he cut too many trees, the hole might be too large to seed in for decades or longer. Too few trees, and the wall of trees would compete with and suppress seedlings in the small opening, preventing regeneration. The camp's five-year management plan suggested removing groups totaling no more than twenty trees, but Platt was discovering that he might need to take more. "Next go round, we'll increase the size of the opening so we can get fairly dense patches of young trees in there," he said.

Platt's flexible approach to his management activities is known as "adaptive management." This approach considers management as a learning process, a series of experiments. Managers constantly monitor results so they can make small course adjustments to avoid management catastrophes that might occur by adhering to a rigid plan.[8]

"Jean said that originally she thought we would follow our five-year plan and then reevaluate it," Platt noted. "But what's happened is that we're evaluating all the time, we're always doing something different from what we said we were going to do and that's because we're continually accumulating information. The quicker we can adjust our management, the better off we are. Adaptive management gives you a chance to make a quicker decision without being tempted to scrap the entire management plan, which is what often has happened on federal land."

At Camp Whispering Pines, Bill Platt and his management team were attempting to restore a degraded longleaf pine forest year by year: burning often and aggressively during the growing season, removing off-site pines and privet, and imitating natural processes that open small holes in the canopy to encourage biodiversity and regeneration. The eventual goal was to restore a mature forest ecosystem with trees of all ages, a native fauna, a full array of native plants, and functioning natural processes.

"Already we have seen the native flora and fauna respond, so that there are more blazing stars, more fence lizards, more Bachman's sparrows, and more fox squirrels than a decade ago," Platt said.

The full achievement of the goal is centuries away—most longleaf there were scarcely teenagers, only sixty to seventy years old. But on the camp's best tracts, Platt estimated, a maintenance management phase was within reach in about twelve years. On others, especially where the ground cover was totally missing, restoration practices would need to be in place for forty or fifty years.

—————

"Here's my basic philosophy behind the restoration of longleaf pine," Finis Harris said. "Number one, apply fire. Low-intensity fires are the only thing that will bring back the ground cover. Number two, apply an appropriate harvesting technique. Group selection is the best harvesting approach to restoring the forest and plant communities."

I scrambled to open my notebook as Harris, the forest silviculturist on the giant, 600,000-acre Kisatchie National Forest in central and western Louisiana, made these remarkable statements. We were just minutes into a day of touring the Kisatchie in Harris's pickup, and I had already begun to realize that I was dealing with an unusual U.S. Forest Service employee.

It was not the last time that Harris would surprise me that day. A quiet, soft-spoken man, he had worked with the Forest Service for thirty-four years, and yet his concerns seemed indistinguishable from Bill Platt's. "For so many decades the emphasis in our agency was to be a productive contributor to the national economy through the production of wood fiber," he said. "Now we know that the production of wood fiber is not the most important thing. Forest health is more important. It's a big breakthrough, this thinking. It's a change in our approach to managing public lands." He voiced as much enthusiasm for the ground cover in the forest as the trees—"In forests in public ownership, we need to restore plant communities for the enjoyment of citizens because they may well be the only places they may see and enjoy them." And on the subject of uneven-aged management, long viewed skeptically by the Forest Service, Harris was adamant: "It's the only way you can achieve the attributes of an old-growth forest."

At one point during the day, I cocked my head at him and told him he didn't sound at all like a typical forester.

"I'm out of step," he replied. "I've always been. In the office, they call me 'Ecofreak.'"

The history of the Kisatchie National Forest is similar to that of the Choctawhatchee National Forest and just about every other national forest in longleaf areas of the South. Originally about 85 to 90 percent of Kisatchie's lands were in longleaf pine, but when the timber barons departed in the 1920s, they left behind another vast stump orchard. In some areas, longleaf began to grow back because of a seedfall just before the clear-cut, but on the vast majority of the forest lands no trees grew. Kisatchie's first foresters' initial objective was to reforest the cutover lands, and they chose to do it with loblolly and slash pine. When Harris arrived at Kisatchie in 1993, the existing conditions pretty nearly matched what Bill Platt found at Camp Whispering Pines: loblolly and slash pines growing on two-thirds of the longleaf sites; abundant sweet gums, dogwoods, and scrub oaks because of years of fire suppression; and predominantly even-aged forests in place of the original uneven-aged forests.

Kisatchie's goal is to more than double the amount of longleaf. Harris and his colleagues at Kisatchie have described the national forest's "desired future condition" as "open stands of pure longleaf pine, with few if any midstory hardwood trees, over rich and productive herbaceous plant communities intermixed with understory hardwoods kept in check by repeated prescribed burning." The Kisatchie will look a lot like the Wade Tract, Harris told me. For him, as well as for Bill Platt and so many others across the South, the Wade Tract has become a visual guide to what an old-growth longleaf pine forest should look like.[9]

Fire will be a main tool in the restoration of the longleaf, that and the harvesting of small groups of loblollies and slash pines in patches no larger than two acres. Seedlings will be planted where there is no nearby seed source for longleaf. In areas of solid longleaf, foresters will create canopy openings of less than an acre where longleaf seedlings will regenerate naturally. The result in time will be an uneven-aged forest.

Harris brought me out to a variety of sites in the Kisatchie to examine the progress that had been made so far. His kind of management is an exercise in understanding how the landscape dictates the growth patterns of plant communities and the kind of management necessary to sustain them. For example, he pointed out how plant communities respond to fire differently depending on where they are situated on a slope: the fire intensity decreases as it burns down the slope, but strengthens as it burns higher toward the dry ridge tops.

"That difference in fire intensity creates different plant communities," he said. "You've got one type of plant community on the upland, and a different plant community on the bottom, and even another plant community in the transition zone. That's what causes your mosaic pattern, and that's a pattern on the land you can recognize. You've got diversity within the plant community, and you've got diversity between the plant communities. All of it was driven by fire."

And all of it required different kinds of management. At one uneven-aged site in the Kisatchie District, the foresters had harvested the off-site pines on the uplands, leaving the longleaf. Then they had applied herbicides to the hardwood trees that had grown too large to be controlled by fire. Down in the drains, however, where they grew naturally, the hardwoods were left. "We haven't put a fire through this site sufficiently because I can still see some off-site loblolly," Harris admitted. "I keep trying to get a good growing-season fire and they keep assuring me that they can get it, but they haven't. I can see we're going to need to come in here with cutting tools and remove those young loblollies. What I keep telling folks is that 70 percent of our prescribed burns need to be done in the growing season."

As we walked through the site, he pointed out a spot where a dense and bristling cohort of young longleaf stood, noting with satisfaction that the uneven-aged character of the forest was beginning to emerge. The young trees were growing in openings created by the removal of the off-site pines. Harris realized that the denser the forest canopy, the poorer the community of native plants in the understory. Creating openings was helping to restore the entire ecosystem by allowing native grasses to grow, thus providing more fuel for fires and encouraging the original plants, which would, he hoped, provide habitats for wildlife.

"We're aiming at a diverse ground cover," Harris added. "The more diverse the ground cover, the more diverse the insect population—it's a basic food chain. Once you get your ground cover in place, the red-cockaded woodpecker will fly."

Time was in favor of this kind of management. The longleaf might only be sixty years old, but he figured that the forest had four hundred years in which to grow into an old-growth forest. The natural longevity of the longleaf pine enables a grower to work slowly and patiently, removing groups of trees here and there, establishing new areas of regeneration, cajoling the forest and shaping it, participating in a process that will last far beyond Harris's lifetime.

In his approach to forest management, he reminded me of Leon Neel. "Leon is a master forester," Harris acknowledged. What he admired most about Neel's work was his almost tactile sense of the forest. Harris seemed to have it, too. As the forest silviculturist presiding over 600,000 acres, his responsibilities were many times greater than Neel's, but he was doing it in a similar hands-on way — recommending where to make third-of-an-acre cuts and where to burn more aggressively. I thought how extraordinarily difficult it must be to provide this kind of management on such a large property and how much knowledge and experience it presupposed.

What would happen when he retired, I asked him once that afternoon. Were younger foresters training under him?

He shook his head. "Young people normally pick up what others know, but we don't have young people coming into my area," he replied. "Our funding has been reduced and so has the number of personnel. We're practically at the point where we can't do management, much less worry about people like myself retiring. It's important to pass along this knowledge."

Even worse, he feared, was the lack of encouragement the agency was giving to personnel to remain in a single district long enough to become knowledgeable about the forest type they were managing. His words were reminiscent of something Bill Boyer had told me. "They don't have people staying in one place very long," Boyer said. "So you may have someone there who knows what he's doing and the next thing he's transferred out West somewhere and somebody new comes in from the West." The management of national forests has often been based on calendar schedules to cope with a lack of experienced personnel. Some years ago Boyer took an inspection tour of a southern national forest where a winter burn was scheduled. The previous fall, one of longleaf's infrequent, unexpected seed crops had occurred. Instead of delaying the fire to take advantage of the fallen seed, thus allowing the seedlings time to grow, Forest Service technicians burned on schedule and destroyed the entire seed crop.

Nine years after the longleaf restoration program began at Kisatchie, Harris could only claim an additional three thousand acres of longleaf. It was not an insignificant number, he said, given that in mid-2002, the Kisatchie was just emerging from a lawsuit that had halted its timber cutting, and thus its restoration efforts, for two years. The forest management plans of just about every national forest in the country have run afoul of legal challenges by environmental groups in recent years, although it is in neighboring Texas that court cases have severely limited forest operations for almost twenty years. Starting in 1985, the Texas Committee on Natural Resources (TCONR), joined often by the Sierra Club and the Wilderness Society, filed several suits to challenge the Forest Service's even-aged management practices on the Texas national forests. The earlier suits changed the way longleaf pine was managed in all southern national forests, but many critics have claimed that the later suits have gone too far. In the name of the red-cockaded wood-pecker, these suits have attacked not only even-aged harvesting but also such a widely recognized management practice as prescribed fire. Without pre-scribed fire, biologists, ecologists, and foresters agree, longleaf pine will not survive.[10]

As we returned to the Kisatchie's main office in Pineville, Harris spoke darkly about these problems. The Kisatchie's current forest plan, created in 1999, reduced the levels of timber harvesting, which was frustrating to Harris.

"We need to be harvesting four times what the plan calls for," he said. "Past management here has not been favorable to longleaf restoration, but now that the climate is better we have to take advantage of it to the fullest extent. That's the frustrating thing."

Ironically, the restoration of the longleaf pine ecosystem in national for-ests, a goal long pursued by conservationists, has finally been embraced by the Forest Service only to run into a public that skeptically opposes tree harvesting, even the harvesting of off-site pines to make room for longleaf.

"If we're going to restore longleaf, we absolutely have to be able to cut trees," Harris emphasized. "You can't bias your management by harvesting for wood fiber, but you also can't be biased against cutting trees. There's a middle ground there. But there's one certainty: Without management you will lose what you're trying to achieve."

━━━━━━

Can we really restore something as complex as the longleaf pine ecosystem? Given the earlier failures of longleaf pine forest management, it's difficult to say for sure. Yet if the new management practices at Eglin Air Force Base, Camp Whispering Pines, Kisatchie National Forest, and other places in the longleaf region are indications of a general trend, it is more than possible. One key to successful ecosystem management seems to be a flexible, adap-tive approach that encourages close monitoring and constant assessment to help a manager head off a possible disaster. Another key, and perhaps an even more important one, is acknowledging human fallibility, a step that lies at the heart of adaptive management. Humility in the face of the overwhelming complexity of forested ecosystems may be the first step toward what Reed Noss calls a "cautious and minimally intrusive kind of management."[11]

A Presence on the Land?

You can't imagine how pretty it was down in this country.
—Sedgie Griffith, Mississippi landowner (1995)

A new world is discovered, with an abundance of forests and wildlife, a well-watered land with a profusion of rivers running to the sea. By attempting to exploit the riches of these forests, people nearly destroy them. This part of longleaf pine's story is as old as the nation, perhaps as old as human civilization, a story of greed, wanton waste, and ignorance. But there's another part to longleaf's story, how for more than a century others have struggled to understand the forest and restore it, how they tried and failed and yet continued to try. This part of the longleaf story is a cautionary tale about the limits of science in understanding ecosystems, the complexities of managing them, and the social and political obstacles facing those who seek to restore a damaged ecosystem. This part of longleaf's story is full of hope, but has it come too late?

The longleaf landscape has changed quite a bit since I began my research some years ago, and many of my ideas have changed as well. My assumptions about management, for example, are less skeptical than they were when I began. Then, I might well have said, as many do today, that nature grows forests better than humans do, so the less handling the better. Today I feel more at home among those who believe that humans took on the role of managing forests the moment they set foot on the continent, that we have been managing the longleaf pine forests in one way or another for the last twelve thousand years, and that our responsibilities to manage these forests well are greater today than they have ever been. Although a healthy dose of humility is essential to anyone who attempts to manage nature's productions, I no longer doubt that the work itself is necessary.

Part of this inner change has come about by bearing witness to the doggedness and good faith with which forest managers and researchers have pursued their difficult work. I remember being struck by forester E. V. Roberts's admission in 1931 that he had been unable to regenerate longleaf on the Choctawhatchee National Forest. "Whatever the cause, it is obvious that we have failed to satisfactorily restock the cut-over areas on the Forest," he wrote, "and to this extent have failed in our responsibility as Forest Managers." I admired Roberts's high-minded sense of his vocation as a forester, the forthright way he embraced his responsibilities, his willingness to accept failure. What he and many others have taught me is just how hard the business of understanding forests and managing them is.

Every longleaf advocate in the Southeast has a folder of stories of what is still going wrong in the management of longleaf, especially in national forests. How foresters still seem unable to grasp the simplest ecological needs of the forest — fire, for example. How foresters will destroy the ground cover to plant longleaf and then wonder why they aren't able to get a good fire to burn. How shelterwood cuts are still being left with too few seed trees. How in many places ecosystem management is being observed with a wink and a nod amid business as usual.

But I have been convinced that management is essential if we want to keep the forests we have, especially the forests of longleaf pine. I listened to the sobering words of Duke University ecologist Norman Christensen at the Tall Timbers Research Station's Seventeenth Fire Ecology Conference in 1989. In the wake of the giant fires in 1988 at Yellowstone National Park and the firestorm over U.S. Park Service policy to let "natural" fires burn, he urged a more realistic conception of wilderness management. "The fallacy is of course in the assertion that wilderness has, can, or should exist totally absent of human activities," he said. "No landscape is free of human impacts. . . . Indeed, we have created a world in which there is no such thing as not managing."[1]

Another change: I was once certain that I would be telling a story about an ecosystem on its way to extinction; indeed, forest survey data, showing an overall decline in longleaf acreage, still seem to confirm that forecast. Longleaf on public lands is stable, but there are broad declines on forest industry land and an even deeper reduction in longleaf acreage on private lands where most longleaf grows. In every state where longleaf once grew, with the sole exception of Mississippi, there is less of it today on properties owned by small landowners than in 1985 — much less. Ecologists have listed longleaf pine among the most seriously endangered ecosystems in the country.[2]

Yet I can't help feeling that something has changed and that the promise of longleaf restoration has undergone a remarkable rejuvenation. One day, not long before I finished this book, I went to lunch with a friend. He had been talking to his forest consultant about land his mother had left him in South Carolina. The land was growing loblollies and had already been thinned once. They were going to clear-cut, he said, and I nodded. That was a traditional pattern among many southern forest owners: thin, clear-cut, and plant loblolly pine seedlings.

He surprised me, however. "He says we ought to plant longleaf," my friend told me. Although a single decision to grow longleaf doesn't necessarily mean better times for the longleaf ecosystem, one hears a lot of these kinds of stories. My friend's decision indicates that if there is plenty of reason for pessimism about the future of longleaf on private lands, anecdotally, at least, there is also some room for hope. Whereas once forestry consultants routinely warned their clients away from longleaf, many are now suggesting that it can very well make them money. Whereas once a private landowner would have dismissed the idea of growing longleaf—*I'd be a fool to plant longleaf*—there are many who are doing just that.

I talked with Rhett Johnson and Dean Gjerstad, codirectors of the Longleaf Alliance, about how to reconcile what seems to be the growing enthusiasm by private landowners about longleaf with data that show a continued decline in longleaf acreage. Since its founding in 1995, the alliance has served as a clearinghouse for information on longleaf, holding workshops and conferences for landowners, and lobbying for legislation that would favor longleaf pine. Johnson said he believed that the resurgence was real, but that it hadn't shown up in the data yet. "I would be afraid to say we will reverse the trend," he told me, "but I believe we have halted the decline and maybe made some incremental gains."

Gjerstad was cautiously optimistic, pointing out that the people that he and Johnson generally see are at one end of the spectrum of opinion about longleaf, and that there are still a lot of landowners cutting out their longleaf in favor of the faster-growing loblolly. He was much more hopeful about the effects of cost-share programs on longleaf, especially the Conservation Reserve Program (CRP). The CRP is a voluntary program that provides cost-share payments to farmers who reduce their cropland acreage by planting long-term conservation covers such as grass or trees. In 1998 longleaf pine was named a National Conservation Priority Area, meaning that farmers in the longleaf belt could be eligible for cost-share payments if they grew longleaf on former cropland. The effects on longleaf planting were almost instantaneous, Gjerstad observed. In 1996 Longleaf Alliance figures showed 61 million longleaf seedlings sold and planted; in 2000 the number almost doubled to 115 million seedlings, equal to as many as 210,000 acres planted in longleaf. Of course, there is no telling how many acres of longleaf were lost during the same period, and by themselves longleaf plantations do not indicate a thriving ecosystem.

"It'll take a long time to change attitudes," Gjerstad said. "Still, I think there is hope for real change."

Managers today can avail themselves of a solid foundation of ecological and silvicultural knowledge that they did not possess even twenty years ago, but will this knowledge alone be enough to reverse the tide of longleaf losses? I don't think so, although such a reversal is impossible without it. The key question is no longer whether forest managers understand longleaf enough to restore it. That question has been answered and answered in the affirmative. The key question is, Do we have the will to restore longleaf? If we do, the future is full of challenges.

Foremost of these challenges is the increasing population growth in the South. Over the last twenty years, the South has grown faster than any other region of the country, and one of the hot spots of urban growth is the Atlantic and Gulf Coastal Plains where most of the remnant longleaf pine is located. The 2002 Southern Forest Resource Assessment predicted that urban areas in the South would increase from about 20 million acres in 1992 to 55 million acres in 2020, and to 81 million acres in 2040. This spiraling development threatens to hasten the longtime conversion of natural forest land to residential and commercial development, and to pine plantations. In fact, by the end of the twentieth century the acreage of plantation pine in the South equaled that of natural pine and will no doubt exceed it in the future.[3]

Population growth also means a continuing proliferation of roads and suburban development, one of the chief reasons for the fragmentation of the landscape and the loss of ecological integrity. Roads and development also bring people closer to public lands that burn and into conflict over the ecological needs of forests for fire. People who move from the urban North or South into close proximity of combustible forests frequently misunderstand the needs of a fire-managed ecosystem.

One of the major causes of conflict is the smoke created by fire. Smoke is a safety hazard, sometimes causing car accidents when it blankets a road. It's also a pollutant, containing several particulates that can trigger or worsen respiratory ailments in people, especially the elderly. Strict smoke management guidelines help forest managers decide when and if to burn, but liability issues have increased the number of lawsuits over fire and smoke. Despite laws such as Florida's Prescribed Burning Act (1990), which protects state-certified landowners from liability in case of an escaped fire or smoke from a fire, the threat of liability is effectively inhibiting the use of prescribed fire. Eliminating prescribed fires is a double-edged sword in fire-prone forests because it increases the amount of combustible fuel on the ground as well as the severity of any resulting fire. As the district ranger of the longleaf pine–rich Croatan National Forest in North Carolina told me, "The Croatan will burn. It will either burn by prescription, under a human drip torch, or it will burn catastrophically."[4]

Another serious challenge facing longleaf restoration is the nation's growing reliance on the South as its wood basket. Reduced timber harvests in the Pacific region have shifted wood production southward; in 1997 the South produced 58 percent of the nation's wood fiber and nearly 16 percent of the world's. This attention will increase pressure on all southern forests, especially on the privately owned longleaf forests. Increasing timber prices give landowners more incentive to harvest their forests.[5]

When William Bartram traveled through the Southeast, he was within sight of grand forests of longleaf pine almost everywhere he went. We will never be able to experience an open, parklike forest stretching for hundreds of miles, as Bartram did, nor will we ever be bored by the endless monotony of a landscape of pines. That excess and immensity are gone. The generous redundancy of the "Land of the Longleaf Pine," like that of the tallgrass prairie or the buffalo, has been reduced and simplified.

Yet in almost every state where longleaf once grew, you can still experience something of what Bartram saw—within North Carolina's Fort Bragg Military Reservation in Southern Pines, where regular fires now keep beautiful expanses of wiregrass and longleaf open; in the Red Hills region of Georgia, where you drive past rolling country where thousands of acres of longleaf grow; in the Apalachicola National Forest in Florida, with its miles of open flatwoods; in the Sandhill Crane National Wildlife Refuge in Mississippi, with its wet savannas and prodigal flowering of pitcher plants. Such lands are where you will find the ecosystem most intact, with a flourishing ground cover and a native suite of wildlife species. Such large chunks of longleaf are fragmented and diminished for sure, but increasingly well managed and, visually, still capable of stirring in the visitor the same powerful emotions experienced by Bartram and the others who described the forest when it was whole.

As for the future of longleaf on private lands, you never know. One thing I have learned about longleaf is not to underestimate its power to stir people's imaginations. I remember one landowner in the North Carolina Sandhills telling me he wanted to restore his one hundred acres of longleaf to "presettlement conditions." He had learned the term after attending a couple of the annual conferences devoted to longleaf pine that were held in North Carolina in the 1980s and 1990s. That's where he had first seen slides of oak-choked longleaf pine forests in Florida gradually morphing into open, park-like stands after several cycles of spring burning. It had been a powerful experience, and when he returned he had looked at his land in a new way.

"When I say presettlement times, I'm thinking about the way it was back 500 years or even 2,000 years," he said. "I see a savanna with pines. Open forests."

Decisions like these are possible signs of a new way of thinking about this beleaguered native forest. And one decision at a time, it's surely the only way that longleaf will ever become a significant presence on the land once again.

Notes

PROLOGUE

1. U.S. Forest Service, "Southern Research Station Employees Honored"; Frost, "Four Centuries," 20; Owsley, *Plain Folk*, 39–40.
2. Ware, Frost, and Doerr, "Southern Mixed Hardwood Forest," 463; Outcalt and Sheffield, "Longleaf Pine Forest," 2; BBC News Online, "Amazon Forest"; Reed F. Noss, "Longleaf Pine and Wiregrass"; Norse, *Ancient Forests*, 6.
3. Noss, LaRoe, and Scott, "Endangered Ecosystems," 13; Frost, "Four Centuries," 17.
4. Defebaugh, "Relation of Forestry to Lumbering," 151.
5. Landers, Van Lear, and Boyer, "Longleaf Pine Forests," 39; Percival Perry, "Naval-Stores Industry in the Old South, 1790–1860," 526.
6. Ivy, *Long Leaf Pine*, 4.
7. Wells and Shunk, "Vegetation and Habitat Factors," 487.

CHAPTER I

1. William Bartram, *Travels*, 110.
2. John Bartram, "Diary of a Journey," 43 (warm morning); Peter Kalm, in ibid., 3–4 (did not care to write). Biographical information about William and John Bartram is drawn from Fagin, *William Bartram*; Slaughter, *The Natures of John and William Bartram*; and William Bartram, *Travels*.
3. Quotations in this and the next five paragraphs are from William Bartram, *Travels*. Bartram's journey from Alachua Savanna to Talahasochte begins on p. 137.
4. Hall, *Travels in North America*, 3:256.
5. Claiborne, "A Trip through the Piney Woods," 514; McDaniel, "Vegetation of the Piney Woods," 174.
6. Muir, *Thousand-Mile Walk*, 35; Schwarz, *Longleaf Pine*, 101–2.
7. Oglethorpe to the Trustees; Gordon, "Journal," 16.
8. "Extract of the Journals of Mr. Commissary Von Reck," 48–49 (not to go into the Woods); Whitefield, "Journal of a Voyage from London," 295 (Two men who disappeared); Anonymous letter, in Mills Lane, *Oglethorpe's Georgia*, 1:40 (In Purysburg).
9. Schoepf, *Travels in the Confederation*, 103; Schaw, *Journal*, 141; Goff, "Great Pine Barrens," 24 (Kemble quotation); Alvarez, *Travel on Southern Antebellum Railroads*, 153 (Mackay poem).

10. *Mobile Daily Register*, January 28, 1883.

11. Clark, *South Carolina*, 248 (McKay quotation); Lieberman and Goolrick, "Naval Stores Ages-Old," 32-A (Eugene L. Schwaab quotation, 1853); Crayon, *Old South Illustrated*, 188.

CHAPTER 2

1. Stout and Marion, "Pine Flatwoods," 376 (forty-two major streams); Ware, Frost, and Doerr, "Southern Mixed Hardwood Forest," 463.

2. On Coastal Plain geology, see Walker and Coleman, "Atlantic and Gulf Coastal Plain"; Martin and Boyce, "Introduction: The Southeastern Setting," 18–22; Stout and Marion, "Pine Flatwoods," 377–79; Clark and Stearn, *Geological Evolution of North America*; Fenneman, *Physiography of Eastern United States*; Matsch, *North America and the Great Ice Age*; and Richards and Hudson, "Atlantic Coastal Plain."

3. Paul Delcourt et al., "History, Evolution," 62.

4. Cabeza de Vaca, *Castaways*, 20; Romans, *Concise Natural History*, 90; Claiborne, "Rough Riding Down South," 37.

5. Komarek, "Natural History of Lightning."

6. Chen and Gerber, "Climate," 26 (thunderstorm days); Komarek, "Fire Ecology — Grasslands and Man," 174; Sharon Hermann, personal communication (a log at the Wade Tract smoldered).

7. Wahlenberg, *Longleaf Pine*, 3–6.

8. Komarek, "Fire Ecology" (1962), 96.

9. Hall, *Travels in North America*, 3:250–51.

10. Pyne, *Fire in America*, 155; Bruce Means, personal communication (fire in Albany); Means and Grow, "Endangered Longleaf Pine Community," 3 (most longleaf pine forests would have burned).

11. Hare, "Contribution of Bark to Fire Resistance."

12. Wahlenberg, *Longleaf Pine*, 72, 119–20; Mirov and Hasbrouck, *Story of Pines*, 39.

13. On longleaf pine seeds, the seeding process, and its relation to fire, see Wahlenberg, *Longleaf Pine*; Wells and Shunk, "Vegetation and Habitat Factors"; Boyer and White, "Natural Regeneration of Longleaf Pine"; Boyer, "Regenerating Longleaf Pine"; Robbins and Myers, "Seasonal Effects"; Christensen, "Fire Regimes"; and Myers, "Scrub and High Pine."

14. For longleaf pine growth strategy, see Wahlenberg, *Longleaf Pine*, 70–118, 95 (taproots can be eight feet long), and Means and Grow, "Endangered Longleaf Pine Community" 2–3.

15. Wahlenberg, *Longleaf Pine*, 115.

16. Ibid., 79.

17. Boyer, "Longleaf Pine Seed Predators"; Wahlenberg, *Longleaf Pine*, 177–83.

18. Janzen, "Seed Predation by Animals"; Silvertown, "Evolutionary Ecology."

19. Ware, Frost, and Doerr, "Southern Mixed Hardwood Forest," 453.

20. The literature on the adaptations of understory plants to fire is voluminous. I have relied especially on Barry, *Natural Vegetation*; B. W. Wells, *Natural Gardens* and "Ecological Problems"; Wells and Shunk, "Vegetation and Habitat Factors"; Robbins and Myers, "Seasonal Effects"; and Christensen, "Fire Regimes" and "The Xeric Sandhills."

21. The seasonal effects of fires in longleaf pine forests have a sizable bibliography. I am indebted to discussions in Robbins and Myers, "Seasonal Effects"; Davis, "Effect of Season of Fire"; and Platt, Evans, and Davis, "Effects of Fire Season."

22. Chen and Gerber, "Climate," 26.

23. Mutch, "Wildland Fires and Ecosystems." On the subject of how frequent fire shaped the traits of longleaf pine and its plant associates, see Christensen, "The Xeric Sandhills"; Platt, Evans, and Davis, "Effects of Fire Season"; Platt, Glitzenstein, and Streng, "Evaluating Pyrogenicity"; Landers, Byrd, and Komarek, "A Holistic Approach"; Ware, Frost, and Doerr, "Southern Mixed Hardwood Forest"; Robbins and Myers, "Seasonal Effects"; and Frost, Walker, and Peet, "Fire-Dependent Savannas and Prairies."

24. Snyder, "'Mutch Ado about Nothing'"; B. W. Wells, *Natural Gardens*, 69.

25. Davis, "Effect of Season of Fire," 43; Platt, Evans, and Rathbun, "Population Dynamics of a Long-Lived Conifer," 516–17.

26. Bendell, "Effects of Fire on Birds and Mammals," 105.

27. Tanner, *Ivory-Billed Woodpecker*, 14; Stoddard, *Memoirs*, 39.

CHAPTER 3

1. B. W. Wells, *Natural Gardens*; Wells and Shunk, "Vegetation and Habitat Factors," 468.

2. Schafale and Weakley, *Classification of the Natural Communities*; Peet, personal communication.

3. B. W. Wells, *Natural Gardens*; Mirov, *The Genus* Pinus, 437. On the droughty conditions of these deep sandy communities, see Wells and Shunk, "Vegetation and Habitat Factors," and B. W. Wells, "Ecological Problems"; see also Barry, *Natural Vegetation*, 103–12.

4. Barry, *Natural Vegetation*, 103–12; Abrahamson and Hartnett, "Pine Flatwoods and Dry Prairies," 106.

5. B. W. Wells, *Natural Gardens*, 79–108.

6. Wilson, *Diversity of Life*, 197; Todd Engstrom, personal communication (nearly 400 plants).

7. Walker and Peet, "Composition and Species Diversity," 163.

8. Peet and Allard, "Longleaf Pine Vegetation," 60 (North Carolina and Gulf Coast figures); Peet et al., "Mechanisms of Co-Existence" (South American figures); Peet, personal communication (other ecologists).

9. Weakley and Sorrie, personal communication.

10. Peet and Allard, "Longleaf Pine Vegetation," 62.
11. Joan L. Walker, "Rare Vascular Plant Taxa," 107.
12. Ruffin, *Sketches of Lower North Carolina*, 254.

CHAPTER 4

1. For general information about the gopher tortoise, I am indebted to Diemer, "Ecology and Management of the Gopher Tortoise" and "Home Range and Movements of the Tortoise," "Tortoise Relocation in Florida," and "Demography of the Tortoise"; Diemer et al., *Gopher Tortoise Relocation Symposium*; Kaczor and Hartnett, "Gopher Tortoise . . . Effects"; Cox, Inkley, and Kautz, "Ecology and Habitat Protection Needs"; and Burke, "Multiple Occupancy."
2. Cox, Inkley, and Kautz, "Ecology and Habitat Protection Needs," 17.
3. On the gopher tortoise's burrow, see Kaczor and Hartnett, "Gopher Tortoise . . . Effects"; Cox, Inkley, and Kautz, "Ecology and Habitat Protection Needs"; Jackson and Milstrey, "The Fauna of Gopher Tortoise Burrows"; and Burke, "Multiple Occupancy," 10 (the record is held).
4. Eisenberg, "The Gopher Tortoise as a Keystone Species"; Jackson and Milstrey, "The Fauna of Gopher Tortoise Burrows," 87 (Sixty vertebrate species and 302 invertebrates); Burke, "Multiple Occupancy," 13 (cottontail rabbit and a six-lined race runner lizard; Feeding on the tortoise feces); Cox, Inkley, and Kautz, "Ecology and Habitat Protection Needs," 13 (Other species that . . . use it for refuge).
5. Kaczor and Hartnett, "Gopher Tortoise . . . Effects," 107; Diemer, "Ecology and Management of the Gopher Tortoise," 126–27; Cox, Inkley, and Kautz, "Ecology and Habitat Protection Needs," 21.
6. Diemer, "Gopher Tortoise Status and Harvest Impact Determination," 25.
7. Jackson, "Red-Cockaded Woodpecker," 5 (*Picus querulus*). The life history of the red-cockaded woodpecker is particularly well studied and documented. I've relied most heavily on Hooper, Robinson, and Jackson, "Red-Cockaded Woodpecker"; Jackson, "Red-Cockaded Woodpecker"; Kulhavy, Hooper, and Costa, *Red-Cockaded Woodpecker*; Lennartz, "Red-Cockaded Woodpecker"; Ligon, Stacey, Conner, Bock, and Adkisson, "Report of the American Ornithologists' Union Committee"; McFarlane, *A Stillness in the Pines*; Jeffrey R. Walters, "Application of Ecological Principles"; Walters, Doerr, and Carter, "Cooperative Breeding System."
8. Ligon et al., "Report of the American Ornithologists' Union Committee," 849 (oldest and largest trees). In Lee and Parnell, "Endangered, Threatened, and Rare Fauna," Jeffrey R. Walters reports that woodpeckers prefer cavity trees 150 to 200 years old in the North Carolina Sandhills (p. 20).
9. Rudolph, Conner, and Turner, "Competition for Red-Cockaded Woodpecker Roost and Nest Cavities," 34.
10. Sinclair, Lyon, and Johnson, *Diseases of Trees and Shrubs*, 350; Hooper, Lennartz, and Muse, "Heart Rot and Cavity Tree Selection," 323, 326.

11. Hooper, Robinson, and Jackson, "Red-Cockaded Woodpecker," 2 (territory); Jackson, "Red-Cockaded Woodpecker," 4 (male foraging).

12. *Final Environmental Impact Statement*, vol. 1, B-1 (family units); Walters, Doerr, and Carter, "Cooperative Breeding System," 276.

13. Walters, Doerr, and Carter, "Cooperative Breeding System," 276; Jeffrey R. Walters, "Application of Ecological Principles."

14. Weigl et al., "Ecology of the Fox Squirrel."

15. Eisner and Wilson, "Conquerors of the Land," 3.

16. Eisner and Wilson, *The Insects*, 2.

17. Ehrlich and Raven, "Butterflies and Plants," 195.

18. Brewer and Winter, *Butterflies and Moths*, 177–78 (flower shapes); Eisner and Wilson, Preface to *The Insects*.

19. Hall's work is reported in Hall and Schweitzer, "Survey of the Moths, Butterflies, and Grasshoppers."

20. Bradshaw and Holzapfel, "Life in a Deathtrap."

21. Frank Morton Jones, "Pitcher-Plant Insects — I," "II," and "III," and "Pitcher-Plants and Their Moths."

CHAPTER 5

1. Watts, "Vegetational History," 301.

2. B. W. Wells, *Natural Gardens*, 90.

3. Michael Williams, *Americans and Their Forests*, 33–34 (Indian populations in the Americas); *Southern Forest Resource Assessment* (Southeastern Indian populations), 599; Hudson, *Knights of Spain*, 12 (language families), and "Genesis of Georgia's Indians," 12 (Indians on the Chattahoochee). For an overall picture of southeastern Indians at the time of European contact, I relied on Ethridge and Hudson, *Transformation*; Hudson, *Knights of Spain*; and Milanich, *Laboring in the Fields*.

4. Clayton, Knight, and Moore, *De Soto Chronicles* (de la Vega, 2:197; "The Account by a Gentleman from Elvas," 1:93).

5. Croker, *Longleaf Pine*, 3; Adair, *History of the American Indians*, 417.

6. "Extract of the Journals of Mr. Commissary Von Reck," 49 (20 or 30 miles); Lyell, *Second Visit to the United States*, 127 (many long rides); James Grant Forbes, *Sketches*, 147 (pine trees were so far from each other).

7. William Bartram, *Travels*; Frost, Walker, and Peet, "Fire-Dependent Savannas," 349 (savanna origin); Ruffin, "Notes of a Steam Journey," 248; Alexander Mackay, "The Western World, or Travels in the United States" (1846–47), in Clark, *South Carolina*, 248–49.

8. Pyne, *Fire in America*, 82.

9. Hudson, *Knights of Spain*, 169 (effects of corn agriculture); Michael Williams, *Americans and Their Forests*, 39 (tallahassees).

10. Kelton, "Great Southeastern Smallpox Epidemic," 36 (decline of 90 to 95 percent); Adair, *History of the American Indians*, 259.

11. Hudson, "Genesis of Georgia's Indians," 12; Milanich, *Laboring in the Fields*, 30; Worth, "Spanish Missions," 52.

12. Hudson, "Genesis of Georgia's Indians," 16; Milanich, *Laboring in the Fields*, 188, 190–91.

13. William Bartram, *Travels*, 118; Romans, *Concise Natural History*, 125 (Arab mounts).

14. Olmsted, *Journey in the Seaboard Slave States*, 348; Ruffin, *Diary*, 182; "The Piney Woods," 368.

15. Claiborne, "Rough Riding Down South," 29; "The Piney Woods," 368.

16. Sitton, *Backwoodsmen*, 198–99; Owsley, *Plain Folk*, 40–41.

17. On cattle ranching traditions in longleaf pine country, see Jordan, *Trails to Texas* and *North American Cattle-Ranching Frontiers*, and McWhiney and McDonald, "Celtic Origins of Southern Herding Practices," 176 (Ireland or Scotland).

18. Schoepf, *Travels in the Confederation*, 108; Gray and Thompson, *History of Agriculture*, 1:149 (1,500 to 6,000 head).

19. Claiborne, "A Trip through the Piney Woods," 520–21; Owsley, *Plain Folk*, 39; Jordan, *Trails to Texas*, 46.

20. Jordan, *North American Cattle-Ranching Frontiers*, 10–11.

21. Frost, "Four Centuries"; Pyne, *Fire in America*, 148; Sitton, *Backwoodsmen*, 202.

22. Schoepf, *Travels in the Confederation*, 110; Kirby, *Pocosin*, 101.

23. Owsley, *Plain Folk*, 50.

CHAPTER 6

1. Cabeza de Vaca, *Castaways*, 28.

2. Blount, *Spirits of Turpentine*, 6; Gamble, *Naval Stores*, 17 (trees which could supply).

3. Torr, *Ancient Ships*, 36 (patches of color). On the classical uses and origins of tar and pitch, see Casson, *Ships and Seamanship*; Morrison, *Greek Oared Ships*; and Torr, *Ancient Ships*.

4. Casson, *Ships and Seamanship*, 194.

5. Morrison, *Greek Oared Ships*, 308; Muir, *Thousand-Mile Walk*, 179.

6. On the uses of tar and pitch in shipbuilding, see Goldenberg, *Shipbuilding in Colonial America*, and Albion, *Square-Riggers on Schedule*.

7. Gamble, *Naval Stores*, 42.

8. Information on ropes and ropemaking is from Margherita Desy, Associate Curator, USS Constitution Museum, personal communication, and Rees, *Rees's Naval Architecture*, 144–47.

9. Brewington, "Documents."

10. "Tables of Allowances of Equipments, Outfits, Stores Etc, for Each Class of Vessels in the Navy of the United States"; App. R, "List of Stores and Outfits for a First-Class

Whale-Ship, for a Cape Horn Voyage," Scammon, *Marine Mammals of the Northwestern Coast*, 313–19.

11. Dana, *Two Years before the Mast*, 50–51.

12. *Compact Edition of the Oxford English Dictionary* (standard naval punishment).

13. *Spirit of Massachusetts*, 28 (troublesome seam); Dickens, *Bleak House*, 216.

14. Burgess, *Coasting Captain*, 23.

15. Gray and Thompson, *History of Agriculture*, 1:154.

16. Ibid., 153–54; Percival Perry, "Naval-Stores Industry in the Old South, 1790–1860," 511; Lefler and Newsome, *North Carolina*, 97 (70 percent of the tar exported).

17. Crittenden, *Commerce*, 56–57; Gray and Thompson, *History of Agriculture*, 1:156.

18. MacLeod, "Tar and Turpentine Business," 17.

19. Cross, "Tar Burning."

20. Eugene L. Schwaab, 1853, quoted in Lieberman and Goolrick, "Naval Stores Ages-Old," 32-A.

21. Catesby, *Catesby's Birds of Colonial America*, 155 (about 180 cords); Silver, *New Face on the Countryside*, 128 (75,000 cords).

22. MacLeod, "Tar and Turpentine Business," 18.

23. Powell, "What's in a Name?"

CHAPTER 7

1. Kirke, *Among the Pines*, 34.

2. Outland, "Servants of the Turpentine Orchards," 20; Percival Perry, "Naval-Stores Industry in the Old South, 1790–1860," 524.

3. Scores of eighteenth- and nineteenth-century travelers and other writers described the curious process of making turpentine and rosin. See, e.g., Averitt, "Turpentining with Slaves"; Gamble, *Naval Stores*, 10–16, 25–27; Herriot, "Manufacture of Turpentine"; MacLeod, "Tar and Turpentine Business"; Olmsted, *Journey in the Seaboard Slave States*, 343–48; "Pine Forests of the South"; Ruffin, "Notes of a Steam Journey," 251; "Southern Pine Forests—Turpentine"; *Tenth Census of the United States, 1880: Forests*, 5415–18; "Turpentine: Hints for Those about to Engage in Its Manufacture"; and "Turpentine Making." Among twentieth-century overviews are Percival Perry, "Naval Stores Industry in the Ante-Bellum South"; Gerry, "Naval Stores Handbook"; Outland, "Servants of the Turpentine Orchards"; Hayes, "General History of the Turpentine Industry"; Sharrer, "Naval Stores"; and Butler, *Treasures of the Longleaf Pines Naval Stores*.

4. On resin and sap, see Mirov and Hasbrouck, *Story of Pines*, 36–37, and Wahlenberg, *Longleaf Pine*, 190.

5. Gerry, "Naval Stores Handbook," 31.

6. Ibid., 22–23; Mirov and Hasbrouck, *Story of Pines*, 9–10.

7. M. Jones to James R. Grist, August 22, 1858, Grist Papers, Duke University, Durham. Courtesy of David Cecelski.

8. Haller, "Sampson of the Terebinthinates," 754.

9. Shelton, *Pines and Pioneers*, 197.

10. Porcher, "Uses of Rosin and Turpentine," 30 (waterproofed their boots); Percival Perry, "Naval Stores Industry in the Ante-Bellum South," 212–15 (candlewicks).

11. Sharrer, "Naval Stores," 266; Wicker, *Miscellaneous Ancient Records*, 499.

12. Sharrer, "Naval Stores," 257.

13. Percival Perry, "Naval Stores Industry in the Ante-Bellum South," 61.

14. Sharrer, "Naval Stores," 258; Schorger and Betts, "Naval Stores Industry," 29.

CHAPTER 8

1. Crittenden, *Commerce*, 15–17; Moore, "Voyage to Georgia," 115.

2. Olmsted, *Journey in the Seaboard Slave States*, 374; Johnson, *Riverboating*, 19; *Wilmington Messenger*, March 4, 20, 1888.

3. *Wilmington Daily Herald*, November 1, 1854.

4. Olmsted, *Journey in the Seaboard Slave States*, 368–69; "Turpentine Making," 348 (Chattahoochee).

5. Ruffin, "Notes of a Steam Journey," 251.

6. Letter to editor, signed "Turpentine," *Weekly Communicator*, May 14, 1852; Percival Perry, "Naval-Stores Industry in the Old South, 1790–1860," 520 (Wilmington and Manchester Rail Road).

7. *Wilmington Star*, January 2, 1878.

8. Hayes, "General History of the Turpentine Industry," 5–6 (Colquitt County); Gamble, *Naval Stores*, 78 (Mississippi distilleries).

9. Herriot, "Manufacture of Turpentine," 453 (7,500 to 9,000 boxes); *Wilmington Messenger*, January 5, 1901.

10. Bond, "Development of the Naval Stores Industry," 194; Gamble, *Naval Stores*, 56–57, 71.

11. *Wilmington Star*, October 26, 1887 (sale of spirits by weight); Gamble, *Naval Stores*, 67 (adulteration).

12. Daniel, *Shadow of Slavery*, 37.

13. Ibid.

14. "Southern Pine Forests — Turpentine," 189 (intelligent class); Ida Bell Williams, *History of Tift County*, 30–31 (wore pistols); Armstrong, "Transformation of Work," 524 (camps were drawing away black labor).

15. *Tifton Gazette*, January 29, 1892.

16. The speeches by Mayor Fletcher and Governor Jennings at the Jacksonville Conference were recorded in "Turpentine Operators' Association Holds Its First Annual Convention," *Florida Times-Union and Citizen*, September 11, 1902.

17. Gamble, *Naval Stores*, 35.

18. *Florida Highways*, 12.

19. Gerry, "The Goose and the Golden Egg," 36.

20. Herriot, "Manufacture of Turpentine"; G. W. Perry, *Treatise on Turpentine Farming*.

21. Olmsted, *Journey in the Seaboard Slave States*, 341.

22. Sargent, "Report on the Forests," 515–18 (Carelessly deep gashes); Ashe, "Forests, Forest Lands," 92 (prostrated long-leaf pines).

23. Unknown source, "Naval Stores" File, New Hanover Museum, Wilmington, N.C.

24. Kirke, *My Southern Friends*, 133.

25. Sargent, "Report on the Forests," 517.

26. Schorger and Betts, "Naval Stores Industry," 40.

27. Ashe, "Forests, Forest Lands," 85–86. Olmsted (*Journey in the Seaboard Slave States*, 346) noted, too, that "a North Carolina turpentine orchard, with the ordinary treatment, lasts 50 years." G. W. Perry (*Treatise on Turpentine Farming*, 54) in Craven County, N.C., wrote that "pines ought to last for twenty-five years with their first boxes."

28. Steuart, "Turpentine and Rosin," 1004–5.

29. Gerry, "Naval Stores Handbook," 43.

30. G. W. Perry, *Treatise on Turpentine Farming*, 58; Gerry, "Naval Stores Handbook," 9.

31. *Wilmington Weekly Star*, February 10, 1888.

32. Gamble, *Naval Stores*, 60.

33. G. W. Perry, *Treatise on Turpentine Farming*, 121; Reed, "Saving the Naval Stores Industry," 171; Schorger and Betts, "Naval Stores Industry," 56–58.

34. For Herty's development of the "Herty" cup, see Reed, "Saving the Naval Stores Industry," and Gamble, *Naval Stores*, 135.

35. On the French method of turpentining, see Ashe, "Forests, Forest Lands," 96–105; Gerry, "Naval Stores Handbook," 88–90; and Gamble, *Naval Stores*, 159.

36. Reed, "Saving the Naval Stores Industry," 174; *Wilmington Dispatch*, June 4, 1903.

37. *Wilmington Dispatch*, June 4, 1903.

38. Percival Perry, "Naval Stores," 40.

CHAPTER 9

1. Bryant, *Logging*, 33; Spring, "Report of the Lumbering of Loblolly Pine."

2. Mohr, *Timber Pines*, 12–13; Fernow, *Report of the Secretary of Agriculture, 1891*, 212.

3. Sargent, *Sylva of North America*, 11:153–54 (220-year-old specimen); Roth, "Notes on the Structure of the Wood," 144–45 (twenty-five to fifty annual rings); Herman H. Chapman, "A Method of Studying Growth and Yield," 214; Charles S. Chapman, "Working Plan," 38–39; Sherrard, "Working Plan," 42 (a tree 16 inches in diameter); Fernow, *Report of the Secretary of Agriculture, 1891*, 219 (tensile strength); Buttrick, "Commercial Uses of Longleaf Pine," 905 (stronger than steel).

4. Wahlenberg, *Longleaf Pine*, 33–34.

5. Gamble, *Naval Stores*, 11.

6. Burgwin to an unidentified correspondent (timbers less than 10 inches square); Bry-

ant, *Lumber*, 409; Defebaugh, *History of the Lumber Industry*, 1:541 (Early southern sawmills could not provide).

7. Hutchins, *American Maritime Industries*, 96–97, 99; Hickman, *Mississippi Harvest*, 279, 102 (governments contracted).

8. Cain, *Four Centuries on the Pascagoula*, 144–45; William Bartram, *Travels*, 198.

9. Bryant, *Lumber*, 378–79 (timber raft in the Carolinas). For more insights into lumber rafting in the Southeast, see Cain, *Four Centuries on the Pascagoula*; Eisterhold, "Savannah"; Hickman, "Logging and Rafting"; and Stokes, "Log-Rafting in Louisiana."

10. Schaw, *Journal*, 184–85.

11. Dixon, *Dixon Legend*, 23–24.

12. Michael Williams, *Americans and Their Forests*, 167 (pit sawing); Napier, *Lower Pearl River's Piney Woods*, 74 (circular saw).

13. Eisterhold, "Colonial Beginnings," 151; Merrens, *Colonial North Carolina*, 93; "Journal of a French Traveller," 735; Defebaugh, *History of the Lumber Industry*, 1:501–2.

14. Newton, "Sawmills and Related Structures," 155–58; *Carolina Cultivator* 1, no. 11 (June 1856): 348; Hickman, "Logging and Rafting," 165; Cox et al., *This Well-Wooded Land*, 68.

15. Defebaugh, *History of the Lumber Industry*, 1:iii; Whelchel, "Lumber and Timber Products," 618; Maxwell and Baker, *Sawdust Empire*, 21.

16. Sargent, "Report on the Forests of North America," 517–18; Gamble, *Naval Stores*, 27; Fernow, *Report of the Secretary of Agriculture, 1891*, 330.

17. Whitney, *Coastal Wilderness to Fruited Plain*, 178; Robert W. Wells, "Daylight in the Swamp!," 240.

18. Michael Williams, *Americans and Their Forests*, 238; Gates, "Federal Land Policy in the South," 304; Napier, *Lower Pearl River's Piney Woods*, 74; Sitton and Conrad, *Nameless Towns*, 11; Cowdrey, *This Land, This South*, 112 (Michigan men).

19. *New York Lumber Trade Journal*, May 16, 1887.

20. Gates, "Federal land Policy in the South," 330 (5.7 million acres); Michael Williams, *Americans and Their Forests*, 263–64; Cowdrey, *This Land, This South*, 112; Mohr, *Timber Pines*, 29.

21. Mayor, *Southern Timberman*, 14.

22. Cain, *Four Centuries on the Pascagoula*, 144–45; Napier, *Lower Pearl River's Piney Woods*, 128.

23. Bryant, *Logging*, 214; Napier, *Lower Pearl River's Piney Woods*, 133–34; Hickman, *Mississippi Harvest*, 166.

24. Bryant, *Lumber*, 409.

25. *American Lumberman*, January 7, 1899, 23 (timber material par excellence); Mohr, "Present Condition of Forests," 146 (port shipments); Frost, "Four Centuries," 34 (fifty years).

26. Millet, "Lumber Industry of 'Imperial' Calcasieu," 52.

27. Ibid.

28. Herman H. Chapman, "Why the Town of McNary Moved."

29. Sitton and Conrad, *Nameless Towns*, 139, 131–32.

30. Michael Williams, *Americans and Their Forests*, 263.

31. Hickman, "Yellow Pine Industries," 84–85; Mattoon, "Longleaf Pine," 7; Michael Williams, *Americans and Their Forests*, 262–63, 272; "Crowell Saw Mill Historic District," secs. 8, 1.

32. Reginald D. Forbes, "Timber Growing," 10; Michael Williams, *Americans and Their Forests*, 238 (30 million acres).

33. Reginald D. Forbes, "Passing of the Piney Woods," 185. For the effects of ad valorem taxes on southern forests, see also Michael Williams, *Americans and Their Forests*, 279–80; Steen, "The Piney Woods: A National Perspective," 9; Mayor, *Southern Timberman*, 35–36; and Napier, *Lower Pearl River's Piney Woods*, 140–41.

34. Sherrard, "Working Plan," 22 (high stumps); Charles S. Chapman, "Working Plan," 54 (5.5 million board feet); Bryant, "Close Utilization of Timber," 28, 30.

35. Dixon, *Dixon Legend*, 34.

36. Leonard Slade, Purvis, Miss., personal communication.

37. Cox et al., *This Well-Wooded Land*, 67 (wobbly blade); Maxwell and Baker, *Sawdust Empire*, 19 (sawdust-making machines); Mohr, *Timber Pines*, 47; Sargent, *Silva of North America*, 154 (To construct a single mile); Buttrick, "Commercial Uses of Longleaf Pine," 907 (34 million yellow pine ties); Hough, *Report upon Forestry*, 1:114–18 (trainmen sometimes burned heart pine logs).

38. *Mobile Daily Register*, January 10, 1883.

39. Fickle, *New South*, 8; Bryant, *Lumber*, 306–7, 329.

40. Gamble, *Naval Stores*, 61 (advancing wave of humanity); Defebaugh, "Relation of Forestry to Lumbering," 151.

41. Saley, "Relation of Forestry to the Lumbering Industry," 148.

42. Mayor, *Southern Timberman*, 123–24 (It hurt my eyes); Armstrong, "Transformation of Work," 523 (In 1864); Richardson, *History of Aberdeen*, 45 (when I ride along the new three lane highway).

43. B. W. Wells, *Natural Gardens*, 67.

CHAPTER 10

1. Fernow, "Timber as a Crop," 142. For the assumptions of early natural resource management, see Noss and Cooperrider, *Saving Nature's Legacy*, 72–74.

2. Fernow, "The Forest as a National Resource," 45, 37.

3. The cheerleading role of early government foresters is based on Fernow, "The Forest as a National Resource," "Forest Conditions and Forestry Problems," and "Timber as a Crop"; and Graves, "Present Condition of American Silviculture." See also Michael Williams, *Americans and Their Forests*, 411–24.

4. Ashe, "Longleaf Pine and Its Struggle for Existence," 8; Mohr, *Timber Pines*, 66; Graves, "Present Condition of American Silviculture," 31.

5. Eldredge and Recknagel, "Management of Longleaf Pine," 3. Compare with tables in Sherrard, "Working Plan," 42, and Herman H. Chapman, "A Method of Studying Growth and Yield," 214.

6. Eldredge, "Silvical Report, Florida National Forest"; Eldredge and Recknagel, "Management of Longleaf Pine."

7. Recknagel, "Certain Limitations of Forest Management," 229–30.

8. Roberts, "Management Plan," 20.

9. Fernow, "The Forest as a National Resource," 46; Mohr, *Timber Pines*, 57, 65.

10. Bryant, "Some Notes on the Yellow Pine Forests," 83.

11. Herman H. Chapman, "A Method of Studying Growth and Yield."

12. Bryant, "Some Notes on the Yellow Pine Forests," 77 (reluctant to abandon); Mattoon, "Longleaf Pine," 46 (seed tree law); Herman H. Chapman, "Factors Determining Natural Reproduction," 34.

13. Greeley, *Forests and Men*, 124.

14. Fernow, in Mohr, *Timber Pines*, 13.

15. Ruffin, *Sketches of Lower North Carolina*, 264–72; Mohr, *Timber Pines*, 120; Hough, *Report upon Forestry*, 471; Bryant, *Logging*, 8.

16. Ashe, "Loblolly or North Carolina Pine," 12.

17. Mattoon, "Slash Pine," 411.

18. Ibid., 412, 405.

19. Fernow, in Mohr, *Timber Pines*, 14 (the superior tree), 24 (the standard had declined); Curtis, "Woody Plants of North Carolina," 40; Hough, *Report upon Forestry*, 471 (the least valuable); Brown, *American Lumber Industry*, 166 (poorest of the southern pines); Sherrard, "Working Plan," 5 (eagerly accepted as substitutes); Mohr, *Timber Pines*, 120 (demand . . . was growing); *American Lumberman*, April 27, 1907, 60–61.

20. Reginald D. Forbes, "Timber Growing," 8.

21. Smith, *History of Papermaking*, 395.

22. Cary, "Southern Timber Resources."

23. Oden, "Herty and the Birth of Southern Newsprint," 82.

24. "Southward the Paper-Making Industry Moves," *Manufacturers Record*.

25. Herty, "White Paper from Young Southern Pines," 23.

26. "Slash Pine Enters Lists," *Savannah Morning News*, April 1, 1930 (commonest, worthlessest variety); "New Source of Wealth in South's Pine Forests," 14 (developing an appreciation).

27. Nonnemacher, *Impacts of Forest Industries*, 8.

28. Smith, *History of Papermaking*, 440 (2 billion pine seedlings); *Southern Forest Resource Assessment*, 360 (7 million acres).

CHAPTER II

1. Gatewood et al., *Comprehensive Study*, 4–10.

2. Ryan, *Forgotten Plague*, 8.

3. Speir, *Going South*, 256 (quotation). On tuberculosis as a disease, see ibid.; Ryan, *Forgotten Plague*; Rosenkrantz, *From Consumption to Tuberculosis*; and Jones, *Health-Seekers in the Southwest*.

4. Forry, *Climate of the United States*, 250.

5. For the early history of Thomasville as a health resort, and for the quotations by Dr. T. S. Hopkins and others, see *Thomasville (Among the Pines)*, 1888, and *Thomasville, Georgia: The Great Winter Resort among the Pines*, 1901.

6. *Thomasville (Among the Pines)*, 27.

7. *Piney Woods Hotel*, 20 (Hippocrates); Schoepf, *Travels in the Confederation*, 114.

8. *Mobile Daily Register*, January 3, 1883; Wellman, *County of Moore*, 93.

9. Brueckheimer, "Leon County Hunting Plantations," 67–84 (rural retreat); *Piney Woods Hotel*, 3; Mitchell, *Landmarks*, 77–78 (A forest of yellow pine).

10. *Pinehurst and the Village Chapel*, 9; Hood and Phillips, National Historic Landmark Nomination, 75 (explicit prohibitions).

11. Tom Hill, personal communication.

12. On the origins of the Cooperative Quail Study Investigation, see Stoddard, *Bobwhite Quail*, xxii–xxiv, and *Memoirs*, 242–51; Brueckheimer, "Leon County Hunting Plantations," 140–78; and Komarek, "Quest for Ecological Understanding," "Role of the Hunting Plantation," 176, and "History of Prescribed Fire and Controlled Burning."

13. Stoddard, *Memoirs*, 35.

14. Ibid., 177; Komarek, "Role of the Hunting Plantation," 176 (practical outcome).

15. Herman H. Chapman, "Causes and Rate of Decadence," 16 (single hog); Schwarz, *Longleaf Pine*, 64.

16. Frost, "Four Centuries," 33.

17. Mattoon, "Longleaf Pine," 61.

18. Graves, "Protection of Forests from Fire," 7. On the fire controversy in the southern forests, see Herman H. Chapman, "Forest Fires and Forestry" (1912), "Is Longleaf a Climax?" (1932), and "Fire and Pines," (1944); Greene, "The Forest That Fire Made"; Harper, "Historical Notes" and "Defense of Forest Fires"; Pinchot and Ashe, *Timber Trees and Forests*, 154; Stoddard, *Bobwhite Quail*, 401–14, and "Use of Fire in Pine Forests." See also Komarek, "History of Prescribed Fire and Controlled Burning"; Pyne, *Fire in America*, 112–19; Schiff, *Fire and Water*; Michael Williams, *Americans and Their Forests*, 484–87; and Wahlenberg, *Longleaf Pine*, 57–67, and "Effect of Fire and Grazing."

19. Pyne, *Fire in America*, 199 (worst forest fire); Michael Williams, *Americans and Their Forests*, 449, 321.

20. Pyne, *Fire in America*, 100–101; Stoddard, *Memoirs*, 244.

21. Clark, *Greening of the South*, 49–50.

22. Beadel, "Fire Impressions," 4.

23. Pyne, *Fire in America*, 195–96.

24. Stoddard, *Bobwhite Quail*, 12.

25. Stoddard, *Memoirs*, 244.

26. Stoddard, *Bobwhite Quail*, 401, and "Use of Fire in Pine Forests," 36.

27. Landers, Byrd, and Komarek, "A Holistic Approach," 158; Stoddard, *Bobwhite Quail*, 410, 403.

28. Stoddard, *Memoirs*, 243.

29. Long, "Notes of Some of the Forest Features," 39; Harper, "Defense of Forest Fires," 208.

30. Herman H. Chapman, "Forest Fires and Forestry," 513.

31. Greene, "The Forest That Fire Made," 618, 583.

32. Schiff, *Fire and Water*, 30; Greene, "The Forest That Fire Made," 583.

33. Demmon, "Silvicultural Aspects," 329.

34. Pyne, *Fire in America*, 143.

35. Herman H. Chapman, "Causes and Rate of Decadence," 16; Kirby, *Poquosin*, 216 (they'd set your woods on fire); Maxwell and Baker, *Sawdust Empire*, 67-68.

CHAPTER 12

1. Boyer, "Regenerating Longleaf Pine," 300.

2. Boyer, personal communication.

3. Wahlenberg, *Longleaf Pine*, 100.

4. On the development of the shelterwood method of regenerating longleaf pine, see Croker, "Can the Shelterwood Method Successfully Regenerate Longleaf Pine?," and Boyer, "Longleaf Pine Regeneration and Management."

5. Boyer and White, "Natural Regeneration," 100.

6. Ibid., 99.

7. "Technology Transfer Plan: Longleaf Pine Management," 1, 3.

8. Boyer, "Regenerating Longleaf Pine," 300; Outcalt and Sheffield, "Longleaf Pine Forest," 2.

9. *Southern Forest Resource Assessment*, 374.

10. Bouma, "Alabama: The Big Tree Farm."

11. Ashe, "Forests, Forest Lands," 68; Mattoon, "Longleaf Pine," 22; Wahlenberg, *Longleaf Pine*, 24-25.

12. Rick Hamilton, North Carolina Forestry Extension, personal communication.

13. Schafale and Weakley, "Ecological Concerns," 221.

CHAPTER 13

1. *Sierra Club et al. v. Lyng et al.*

2. Conner, Rudolph, and Walters, *Red-Cockaded Woodpecker*, 77 (population has fallen).

3. Conner, Rudolph, and Walters, *Red-Cockaded Woodpecker*, 222-23 (cull trees), 220-39 (decades of fire suppression); Young and Mustian, *Impacts of National Forests*, 40 (rotations . . . had been shortened).

4. On Hurricane Hugo and the red-cockaded woodpecker population at Francis Mar-

ion National Forest, see Hooper, Watson, and Escano, "Hurricane Hugo's Initial Effects"; Watson et al., "Restoration of the Red-Cockaded Woodpecker Population"; and Hooper and McAdie, "Hurricanes and the Long-Term Management."

5. Hooper and McAdie, "Hurricanes and the Long-Term Management," 3; Hooper, Watson, and Escano, "Hurricane Hugo's Initial Effects."

6. Conner, Rudolph, and Walters, *Red-Cockaded Woodpecker*, 246.

7. The details of regulating an even-aged forest came from Robert Farrar, personal communication.

8. On *Sierra Club v. Lyng*, see McFarlane, *A Stillness in the Pines*, 224–43, and Conner, Rudolph, and Walters, *The Red-Cockaded Woodpecker*, 251–57.

9. Lennartz, "Reply to Edward C. Fritz" (Nature regenerated the longleaf forest); Hall, *Travels in North America*, 3:256; Schwarz, *Longleaf Pine*, 28–29.

10. *Final Environmental Impact Statement*, 1:14–15; Lennartz, "Reply to Edward C. Fritz."

11. Platt, Evans, and Rathbun, "Population Dynamics of a Long-Lived Conifer."

12. Ibid., 510.

13. Engstrom, "Statement of R. Todd Engstrom," 4.

14. Gatewood et al., *Comprehensive Study of . . . the Red Hills Region*, sec. 4, 61.

15. See Farrar, *Fundamentals of Uneven-aged Management in Southern Pine*.

16. U.S. Fish and Wildlife Service, *Recovery Plan for the Red-Cockaded Woodpecker*, 100–103, 198–201.

17. Boyer, "Long-term Development of Regeneration"; Jimmy S. Walker, "Potential Red-Cockaded Woodpecker Habitat."

18. James, Escano, Costa, and Walters, "Panel Discussion: Managed Longleaf Pine Forests," 374.

CHAPTER 14

1. Noss, "Sustainable Forestry or Sustainable Forests?," 22.

2. Frost, *Presettlement Vegetation*, 22; Noss, "Sustainable Forestry or Sustainable Forests?," 22, 27.

3. Joan L., Walker, "Regional Restoration," 8.

4. Johnson and Gjerstad, "Restoring the Longleaf Pine Forest Ecosystem," 18.

5. Pinchot, *Training of a Forester*, 13.

6. Christensen et al., "Report of the Ecological Society of America."

7. Ehrenfeld, *Arrogance of Humanism*, 5 (irrational faith); Kellert and Bormann, "Closing the Circle," 210 (profound shift); Stanley, "Ecosystem Management," 261 (part of nature).

8. For more on the concept of adaptive management, see Carl Walters, *Adaptive Management of Renewable Resources*; Ludwig, Hilborn, and Walters, "Uncertainty, Resource Exploitation, and Conservation;" and Christensen et al., "Report of the Ecological Society of America."

9. Haywood et al., "Protecting and Restoring Longleaf Pine Forests," 133.

10. Conner, Rudolph, and Walters, *Red-cockaded Woodpecker*, 313–14 (gone too far). For some of the early court cases, see McFarlane, *A Stillness in the Pines*, 199–244.

11. Noss and Cooperrider, *Saving Nature's Legacy*, 98.

EPILOGUE

1. Christensen, "Wilderness and High Intensity Fire," 21.

2. Outcalt and Sheffield, "Longleaf Pine Forest."

3. *Southern Forest Resource Assessment*, 160, 163.

4. Brenner and Wade, "Florida's 1990 Prescribed Burning Act."

5. *Southern Forest Resource Assessment*, 306 (58 percent); Outcalt and Sheffield, "Longleaf Pine Forest," 3–4 (especially on the privately owned longleaf forests).

Bibliography

MANUSCRIPT COLLECTIONS

Alabama
Mobile
  Mobile Public Library

Florida
Tallahassee
  Tall Timbers Research Station Library
Valparaiso
  Eglin Air Force Base Archives

Georgia
Savannah
  Georgia Historical Society
Tifton
  Archives of the Georgia Agrirama

North Carolina
Raleigh
  Division of Archives and History
Wilmington
  New Hanover Museum
    "Naval Stores" File
  Underwater Archaeology Office,
    North Carolina Department
    of Cultural Resources
  Bill Reaves Collection
  Wilmington Public Library

PRINTED SOURCES

Abrahamson, Warren G., and David C. Hartnett. "Pine Flatwoods and Dry Prairies." In *Ecosystems of Florida*, edited by Ronald L. Myers and John J. Ewel, 103–49. Orlando: University of Central Florida Press, 1990.

Adair, James. *The History of the American Indians.* 1775. Reprint, New York: Johnson Reprint Corp., 1968.

Albion, Robert Greenhalgh. *Square-Riggers on Schedule: The New York Sailing Packets to England, France, and the Cotton Ports.* 1938. Reprint, Hamden, Conn.: Archon Books, 1965.

Alvarez, Eugene, ed. *Travel on Southern Antebellum Railroads, 1828–1860.* University: University of Alabama Press, 1974.

Armstrong, Thomas F. "Georgia Lumber Laborers, 1880–1917." *Georgia Historical Quarterly* 67 (Winter 1983): 435–50.

———. "The Transformation of Work: Turpentine Workers in Coastal Georgia." *Labor History* 25 (Fall 1984): 518–32.

Ashe, W. W. "The Forests, Forest Lands, and Forest Products of Eastern North Carolina." *North Carolina Geological Survey*, Bulletin No. 5. Raleigh, 1894.

———. "Loblolly or North Carolina Pine." *North Carolina Geological and Economic Survey*, Bulletin No. 24. Raleigh, 1915.

———. "The Longleaf Pine and Its Struggle for Existence." *Journal of the Elisha Mitchell Scientific Society* 11 (January–July 1894): 1–16.

Averitt, James Battle. "Turpentining with Slaves in the 30's and 40's." In *Naval Stores — History, Production, Distribution, and Consumption*, edited by Thomas Gamble, 25–27. Savannah: N.p., n.d.

Barry, John M. *Natural Vegetation of South Carolina*. Columbia: University of South Carolina Press, 1980.

Bartram, John. "Diary of a Journey through the Carolinas, Georgia, and Florida, from July 1, 1765, to April 10, 1766." Edited by Francis Harper. *Transactions of the American Philosophical Society*, n.s., 33 (December 1942): 1–120.

Bartram, William. *The Travels of William Bartram*. 1791. Reprint, Naturalist's Edition, edited by Francis Harper. Athens: University of Georgia, 1998.

BBC News Online. "Amazon Forest 'Could Vanish Fast.'" Monday, June 25, 2001.

Beadel, H. L. "Fire Impressions." *Proceedings of the Tall Timbers Fire Ecology Conference* 1 (1962): 1–6.

Bendell, J. F. "Effects of Fire on Birds and Mammals." In *Fire and Ecosystems*, edited by T. T. Kozlowski and C. E. Ahlgren, 73–138. New York: Academic Press, 1974.

Blount, Robert S. "Spirits of Turpentine: A History of Florida Naval Stores, 1528–1950." *Florida Heritage Journal Monograph*, no. 3. Tallahassee: Florida Agricultural Museum, 1993.

Bond, Stanley C., Jr. "The Development of the Naval Stores Industry in St. Johns County, Florida." *Florida Anthropologist* 40 (September 1987): 187–202.

Bonner, J. C., ed. "Plantation Experiences of a New York Woman." *North Carolina Historical Review* 33 (October 1956): 384–417, 529–46.

Botkin, Daniel B. *Discordant Harmonies: A New Ecology for the Twenty-First Century*. New York: Oxford University Press, 1990.

Bouma, Katherine. "Alabama: The Big Tree Farm." *Montgomery Advertiser*, May 8–14, 1994.

Boyer, William D. "Longleaf Pine Regeneration and Management: An Overstory Overview." *Proceedings of the Longleaf Pine Ecosystem Restoration Symposium*. Society for Ecological Restoration, Ninth Annual International Conference, Longleaf Alliance Report No. 3 (1998): 14–19.

———. "Longleaf Pine Seed Predators in Southwest Alabama." *Journal of Forestry* 62 (July 1964): 481–84.

———. "Long-Term Development of Regeneration under Longleaf Pine Seedtree and Shelterwood Stands." *Southern Journal of Forestry* 17 (February 1993): 10–15.

———. "Regenerating Longleaf Pine with Natural Seeding." *Proceedings of the Tall Timbers Fire Ecology Conference* 18 (1993): 299–309.

Boyer, William D., and John B. White. "Natural Regeneration of Longleaf Pine." In *Proceedings of the Symposium on the Management of Longleaf Pine*, edited by Robert M. Farrar Jr. U.S. Department of Agriculture, Forest Service, General Technical Report SO-75 (1990): 94–113.

Bradshaw, William, and Christina Holzapfel. "Life in a Deathtrap." *Natural History*
(July 1991): 35–36.

Braun, E. Lucy. *Deciduous Forests of Eastern North America*. Philadelphia: Blakiston Co.,
1950.

Brennan, Leonard A., and R. Shane Fuller. "Bobwhites and Red-Cockaded
Woodpeckers: Endangered Species Management Helps Quail, Too!" *Quail
Unlimited Magazine* 12 (May–June 1993): 16–20.

Brenner, Jim, and Dale Wade. "Florida's 1990 Prescribed Burning Act." *Journal of
Forestry* 90 (May 1992): 27–30.

Brewer, Jo, and Dave Winter. *Butterflies and Moths: A Companion to Your Field Guide.*
New York: Prentice Hall Press, 1986.

Brewington, M. V. "Documents." *American Neptune* 2 (April 1966): 141–44, 3 (July
1966): 210–14.

Brown, Nelson Courtlandt. *The American Lumber Industry*. New York: Wiley, 1923.

Brueckheimer, William R. "Leon County Hunting Plantations: An Historical and
Architectural Survey." Final Report: Historical Overview. The Historic Tallahassee
Preservation Board of Trustees, Florida Department of State, 1988.

———. "The Quail Plantations of the Thomasville-Tallahassee-Albany Regions."
Proceedings of the Tall Timbers Ecology and Management Conference 16
(February 22–24, 1979): 141–66.

Bryant, Ralph C. "Close Utilization of Timber." In *Prolonging the Cut of Southern Pine.*
Yale Forest School Bulletin 2 (1913): 23–32.

———. *Logging: The Principles and General Methods of Operation in the United States.*
New York: Wiley, 1913, 1923.

———. *Lumber: Its Manufacture and Distribution.* New York: Wiley, 1922.

———. "Some Notes on the Yellow Pine Forests of Central Alabama." *Proceedings of
the Society of American Forests* 15 (1909): 72–83.

Buell, Murray F. "Late Pleistocene Forests of Southeastern North Carolina." *Torreya*
45 (December 1945): 117–18.

Burgess, Robert H., ed. *Coasting Captain: Journals of Captain Leonard S. Tawes Relating
His Career in Atlantic Coastwise Sailing Craft from 1868 to 1922.* Newport News, Va.:
The Mariners Museum, 1967.

Burgwin, John Fanning. Letter to an unidentified correspondent, 1816. Edited by
Alan D. Watson. In *Lower Cape Fear Historical Society Bulletin* 39 (April 1995).
Newsletter.

Burke, Russell L. "Multiple Occupancy." *Natural History* (June 1992): 8–13.

Butler, Carroll B. *Treasures of the Longleaf Pines Naval Stores.* Shalimar, Fla.: Tarkel
Publishing, 1998.

Buttrick, P. L. "Commercial Uses of Longleaf Pine." *American Forestry* 21 (September
1915): 895–908.

Cabeza de Vaca, Alvar Núñez. *Castaways: The Narrative of Alvar Núñez Cabeza de Vaca.*
Edited by Enrique Pupo-Walker. Translated by Frances M. López-Morillas.
Berkeley: University of California Press, 1993.

Cain, Cyril Edward. *Four Centuries on the Pascagoula: History, Story, and Legend of the Pascagoula River Country*. Vol. 1. State College, Miss.: N.p., 1953.

Cary, Austin. "Southern Timber Resources in Relation to Paper Making," *Paper Trade Journal* (May 8, 1924): 57–60.

Casson, Lionel. *Ships and Seamanship in the Ancient World*. Princeton, N.J.: Princeton University Press, 1971.

Catesby, Mark. *Catesby's Birds of Colonial America*, edited by Alan Feduccia. Chapel Hill: University of North Carolina Press, 1985.

Chapman, Charles S. "A Working Plan for Forest Lands in Berkeley County, South Carolina." U.S. Department of Agriculture, Bureau of Forestry, Bulletin No. 56, 1905.

Chapman, Herman H. "The Causes and Rate of Decadence in Stands of Virgin Longleaf Pine." *Lumber Trade Journal* 84 (September 15, 1923): 1, 16–17.

———. "Factors Determining Natural Reproduction of Longleaf Pine on Cut-over Lands in LaSalle Parish, Louisiana." *Yale University School of Forestry*, Bulletin No. 16, 1926.

———. "Fire and Pines." *American Forests* 50 (February 1944): 62–64, 91–93.

———. "Forest Fires and Forestry in the Southern States." *American Forestry* 18 (August 1912): 510–17.

———. *Forest Management*. New York: J. B. Lyon Co., 1931.

———. "Is the Longleaf Type a Climax?" *Ecology* 13 (October 1932): 328–34.

———. "A Method of Studying Growth and Yield of Longleaf Pine Applied in Tyler County, Texas." *Proceedings of the Society of American Foresters* 4 (1909): 207–20.

———. "Possibilities of a Second Cut." In *Prolonging the Cut of Southern Pine*. Yale Forest School Bulletin No. 2 (1913): 1–21.

———. "Why the Town of McNary Moved." *American Forests* 30 (1924): 589–92, 615–16, 626.

Chen, Ellen, and John F. Gerber, "Climate." In *Ecosystems of Florida*, edited by Ronald L. Myers and John J. Ewel, 11–34. Orlando: University of Central Florida Press, 1990.

Christensen, Norman L. "Fire Regimes in Southeastern Ecosystems." In *Fire Regimes and Ecosystem Properties*, edited by H. A. Mooney, T. M. Bonnickson, N. L. Christensen, J. E. Lotan, and W. A. Reiners, 112–36. U.S. Department of Agriculture, Forest Service, General Technical Report No. WO 26, 1981.

———. "Fire and Soil-Plant Nutrient Relations in a Pine-Wiregrass Savanna on the Coastal Plain of North Carolina." *Oecologia* 32 (1977): 27–44.

———. "Wilderness and High Intensity Fire: How Much Is Enough?" *Proceedings of the Seventeenth Tall Timbers Fire Ecology Conference*, 9–24. Tall Timbers Research Station, 1991.

———. "The Xeric Sandhills and Savanna Ecosystems of the Southeastern Atlantic Coastal Plain U.S.A." *Veroffentlichungen des Geobortanishen Instituts der ETH 68* (1979): 246–62.

Christensen, Norman L., et al., "The Report of the Ecological Society of America

Committee on the Scientific Basis for Ecosystem Management" (1995)
<http://www.sdsc.edu/%7EESA/execsum.htm>.

Claiborne, J. F. H. "Rough Riding Down South." *Harper's New Monthly Magazine* 25
(June 1862): 29–37.

———. "A Trip through the Piney Woods." *Publications of the Mississippi Historical
Society* 9 (1906): 487–538.

Clark, Thomas D. *The Greening of the South: The Recovery of Land and Forest.* Lexington:
University Press of Kentucky, 1984.

———. "The Impact of the Timber Industry on the South." *Mississippi Quarterly* 25
(Spring 1972): 141–64.

———. "The Piney Woods and the Southern Frontier." In *Mississippi's Piney Woods:
A Human Perspective,* edited by Noel Polk, 12–24. Jackson: University Press of
Mississippi, 1986.

———, ed. *South Carolina: The Grand Tour, 1780–1865.* Columbia: University of South
Carolina Press, 1973.

Clark, Thomas H., and Colin W. Stearn. *The Geological Evolution of North America:
A Regional Approach to Historical Geology.* New York: Ronald Press Co., 1960.

Clayton, Lawrence A., Vernon James Knight Jr., and Edward C. Moore, eds. *The
De Soto Chronicles: The Expedition of Hernando de Soto to North America in 1539–1543.*
2 vols. Tuscaloosa: University of Alabama Press, 1993.

The Compact Edition of the Oxford English Dictionary. Glasgow: Oxford University Press,
1971.

Compton, Vernon, and J. J. Bachant. "The Gulf Coastal Plain Ecosystem Partnership:
Fire Learning Network Participating Landscape Site." First Interim Progress
Report, The Nature Conservancy, 2002.

Conner, Richard N. D., Craig Rudolph, and Jeffrey R. Walters. *The Red-Cockaded
Woodpecker: Surviving in a Fire-Maintained Ecosystem.* Austin: University of Texas
Press, 2001.

Cowdrey, Albert E. *This Land, This South: An Environmental History.* Lexington:
University Press of Kentucky, 1983.

Cox, James, Douglas Inkley, and Randy Kautz. "Ecology and Habitat Protection
Needs of Gopher Tortoise (*Gopherus polyphemus*) Populations Found on Lands
Slated for Large-scale Development in Florida." Florida Game and Fresh Water
Fish Commission, 1987.

Cox, Thomas R., Robert S. Maxwell, Phillip Drennon Thomas, and Joseph J. Malone.
This Well-Wooded Land: Americans and Their Forests from Colonial Times to the Present.
Lincoln: University of Nebraska Press, 1985.

Crayon, Porte (David Hunter Strother). *The Old South Illustrated.* Edited by Cecil D.
Eby Jr. Chapel Hill: University of North Carolina Press, 1959.

Crittenden, Charles Christopher. *The Commerce of North Carolina, 1763–1789.* New
Haven: Yale University Press, 1936.

Croker, Thomas. "Can the Shelterwood Method Successfully Regenerate Longleaf
Pine?" *Journal of Forestry* 54 (April 1956): 258–60.

————. *Longleaf Pine: A History of Man and a Forest*. U.S. Department of Agriculture, Forester Service Southern Region, Forestry Report R8-FR7, October 1987.

Cross, John K. "Tar Burning: A Forgotten Art?" *Forests and People* 23 (Second Quarter 1973): 21–23.

"Crowell Saw Mill Historic District, Rapides Parish, LA." Registration form for the National Register of Historic Places, 1993.

Cubbage, Fred, and Don Hodges. "The Economics of Managing Longleaf Pine." In *Proceedings of the Symposium on the Management of Longleaf Pine*, edited by Robert M. Farrar Jr., 215–29. U.S. Department of Agriculture, Forest Service, General Technical Report SO-75 215–229, 1990.

Curran, C. E. "Present and Future Trends in the Pulping of Southern Woods." *Paper Trade Journal* (January 16, 1930): 49–53.

Curtis, Moses Ashley. "Woody Plants of North Carolina (1860)." In *The Woods and Timbers of North Carolina*, edited by P. M. Hale. Raleigh: P. M. Hale Publishers, 1883.

Dana, Richard Henry, Jr. *Two Years before the Mast*. 1840. Reprint, New York: New American Library, 1964.

Daniel, Pete. *The Shadow of Slavery: Peonage in the South, 1901–1969*. Urbana: University of Illinois Press, 1972.

Davis, Mary M. "The Effect of Season of Fire upon Flowering of Forbs in Longleaf Pine-Wiregrass Forests." M.A. thesis, Florida State University, 1985.

Defebaugh, James Elliot. *History of the Lumber Industry of America*. 2 vols. Chicago: American Lumberman, 1906–7.

————. "Relation of Forestry to Lumbering and the Wood-Working Industries." *Proceedings of the American Forestry Association* 10 (1894): 150–58.

Delcourt, Hazel R., and Paul A. Delcourt. "Presettlement Magnolia-Beech Climax of the Gulf Coastal Plain: Quantitative Evidence from the Apalachicola River Bluffs, North Central Florida." *Ecology* 58 (Late Summer 1977): 1085–93.

Delcourt, Paul A. "Goshen Springs: Late Quaternary Vegetation Record for Southern Alabama." *Ecology* 61 (April 1980): 371–86.

Delcourt, Paul A., Hazel R. Delcourt, Dan F. Morse, and Phyllis A. Morse. "History, Evolution, and Organization of Vegetation and Human Culture." In *Biodiversity of the Southeastern United States/Lowland Terrestrial Communities*, edited by William H. Martin, Stephen G. Boyce, and Arthur C. Echternacht, 47–80. New York: Wiley, 1993.

Demmon, E. L. "Silvicultural Aspects of the Forest-Fire Problem." *Journal of Forestry* 33 (March 1935): 323–31.

Description of the New Piney Woods Hotel, Thomasville, Georgia. Reprint, Thomasville: Chamber of Commerce, 1967.

Diemer (Berish), Joan. "Demography of the Tortoise *Gopherus polyphemus* in Northern Florida." *Journal of Herpetology* 26 (September 1992): 281–89.

————. "The Ecology and Management of the Gopher Tortoise in the Southeastern United States." *Herpetologica* 42 (March 1986): 125–33.

————. "Gopher Tortoise Status and Harvest Impact Determination (Final Report)." Florida Game and Fresh Water Fish Commission, 1980–87.

————. "Home Range and Movements of the Tortoise *Gopherus polyphemus* in Northern Florida." *Journal of Herpetology* 26 (June 1992): 158–65.

————. "Tortoise Relocation in Florida: Solution or Problem?" *Proceedings of the Desert Tortoise Council Symposium* (1984): 131–35.

Diemer, Joan E., et al., eds. *Gopher Tortoise Relocation Symposium Proceedings*. State of Florida Game and Fresh Water Fish Commission, Nongame Wildlife Program Technical Report 5 (1989).

Dixon, Solon. *The Dixon Legend*. Huntsville, Ala.: Strode Publishers, 1982.

Ehrenfeld, Paul. *The Arrogance of Humanism*. Oxford: Oxford University Press, 1978, 1981.

Ehrlich, Paul R., and Peter H. Raven. "Butterflies and Plants." In *The Insects*, edited by Thomas Eisner and Edward O. Wilson, 195–202. San Francisco: W. H. Freeman, 1978.

Eisenberg, John. "The Gopher Tortoise as a Keystone Species." In "The Gopher Tortoise: A Keystone Species," edited by Rhoda J. Bryant and Richard Franz, *Proceedings of the Fourth Annual Meeting of the Gopher Tortoise Council*, Valdosta, Ga., 1983.

Eisner, Thomas, and Edward O. Wilson. "General Introduction: The Conquerors of the Land." In *The Insects*, edited by Thomas Eisner and Edward O. Wilson, 2–21. San Francisco: W. H. Freeman, 1978.

Eisterhold, John A. "Colonial Beginnings in the South's Lumber Industry, 1607–1800." *Southern Lumberman* 223 (December 15, 1971): 150–53.

————. "Lumber and Trade in the Lower Mississippi Valley and New Orleans, 1800–1860." *Louisiana History* 13 (Winter 1972): 71–91.

————. "Savannah: Lumber Center of the South Atlantic." *Georgia Historical Quarterly* 57 (Winter 1973): 526–43.

Eldredge, Inman F. "Administrative Problems in Fire Control in the Longleaf-Slash Pine Region of the South." *Journal of Forestry* 33 (March 1935): 342–45.

————. "Silvical Report, Florida National Forest." 1911. Eglin Air Force Base Archives.

Eldredge, Inman F., and A. B. Recknagel. "Management of Longleaf Pine with Special Reference to the Turpentine Industry." Washington, Government Printing Office, 1912. Typescript, courtesy of Eglin Air Force Base.

Emerson, F. V. "The Southern Longleaf Pine Belt." *Geographical Review* 7 (February 1919): 81–90.

Engstrom, R. Todd. "Statement of R. Todd Engstrom, Vertebrate Ecologist, Tall Timbers Research Station, on the Record of Decision and Final Environmental Impact Statement on Management of the Red-Cockaded Woodpecker and Its Habitat on the Southern National Forests," 1995.

Ethridge, Robbie, and Charles Hudson, eds. *Transformation of the Southeastern Indians, 1540–1760*. Jackson: University Press of Mississippi, 2002.

"An Extract of the Journals of Mr. Commissary Von Reck . . . and of the Reverend Mr. Bolzius." In *Our First Visit in America: Early Reports from the Colony of Georgia, 1732–1740*, 41–80. Savannah: Beehive Press, 1974.

Fagin, N. Bryllion. *William Bartram: Interpreter of the American Landscape*. Baltimore: Johns Hopkins Press, 1933.

Farrar, Robert M., Jr. *Fundamentals of Uneven-Aged Management in Southern Pine*, edited by W. K. Moser and L. A. Brennan. Tall Timbers Research Station Miscellaneous Publication No. 9., Tallahassee, Fla., 1996.

———. "Growth and Yield in Naturally Regenerated Longleaf Pine Stands." In *Proceedings of the Tall Timbers Fire Ecology Conference* 18 (1993): 311–36.

———. "A New Look at Longleaf Pine." *The Consultant* (Fall 1991): 18–21.

Fenneman, Nevin M. *Physiography of Eastern United States*. New York: McGraw-Hill, 1938.

Fernow, B. E. "Forest Conditions and Forestry Problems in the United States." *Proceedings of the American Forestry Association* 10 (1894): 29–36.

———. "The Forest as a National Resource." *Proceedings of the American Forestry Association* 7 (1891): 36–53.

———. *Report upon the Forestry Investigations of the U.S. Department of Agriculture, 1877–1898*. 55th Cong., 3rd sess., Document 181. Washington: Government Printing Office, 1899.

———. *Report of the Secretary of Agriculture, 1891*, 191–229. Washington: Government Printing Office, 1892.

———. *Report of the Secretary of Agriculture, 1892*, 293–358. Washington: Government Printing Office, 1892.

———. "Suggestions as to Possibilities of Silviculture in America." *Proceedings of the Society of American Foresters* 11 (April 1916): 171–75.

———. "Timber as a Crop." *Proceedings of the American Forestry Association* 10 (1894): 142–47.

Fickle, James E. *The New South and the "New Competition": Trade Association Development in the Southern Pine Industry*. Urbana: University of Illinois Press, 1980.

Final Environmental Impact Statement for the Management of the Red-Cockaded Woodpecker and Its Habitat on National Forests in the Southern Region. Vols. 1, 2. U.S. Department of Agriculture, Forest Service, Southern Region, Management Bulletin R8-MB 73, June 1995.

Final Environmental Impact Statement, Revised Land and Resource Management Plan, Kisatchie National Forest. U.S. Department of Agriculture, Forest Service, Southern Region, August 1999.

Forbes, James Grant. *Sketches, Historical and Topographical, of the Floridas*. 1821. Reprint, Gainesville: University of Florida Press, 1964.

Forbes, Reginald D. "The Passing of the Piney Woods." *American Forestry* 29 (March 1923): 131–36, 185.

———. "Timber Growing and Logging and Turpentining Practices in the Southern

Pine Region." U.S. Department of Agriculture, Forest Service, Technical Bulletin No. 204, 1930.

Forest Service News Release. "Southern Research Station Employees Honored for Longleaf Pine Ecosystem Restoration Research," June 5, 2000, Southern Research Station, Asheville, N.C.

Forry, Samuel, M.D. *The Climate of the United States and Its Endemic Influences.* New York: H and H. G. Langley, 1842.

Frost, Cecil C. "Four Centuries of Changing Landscape Patterns in the Longleaf Pine Ecosystem." *Proceedings of the Tall Timbers Fire Ecology Conference* 18 (1993): 17–44.

———. *Presettlement Vegetation and Natural Fire Regimes of the Croatan National Forest.* North Carolina Department of Agriculture, Plant Conservation Program, June 25, 1996.

Frost, Cecil C., Joan Walker, and Robert K. Peet. "Fire-Dependent Savannas and Prairies of the Southeast: Original Extent, Preservation Status, and Management Problems." In *Wilderness and Natural Areas in the Eastern United States: A Management Challenge*, 348–57. Nacogdoches, Tex.: Center for Applied Studies, School of Forestry, Stephen F. Austin State University, 1986.

Gamble, Thomas, ed. *Naval Stores — History, Production, Distribution, and Consumption.* Savannah: N.p., n.d.

Gannett, Henry. "The Lumber Industry." In U.S. Bureau of the Census, *Twelfth Census of the United States: Manufacturers* (Part 3), 9:803–997. Special Reports on Selected Industries. Washington: Government Printing Office, 1902.

Garren, Kenneth H. "Effects of Fire on the Vegetation of the Southeastern United States." *Botanical Review* 9 (November 1943): 617–54.

Gates, Paul W. "Federal Land Policy in the South, 1866–1888." *Journal of Southern History* 6 (August 1940): 303–30.

Gatewood, Steve, Kenneth W. Johnson, Richebourg G. McWilliams, Neil G. Sipe, and Nancy Tinker. *A Comprehensive Study of a Portion of the Red Hills Region of Georgia.* Thomasville: Thomas College Press, 1994.

Gerry, Eloise. "The Goose and the Golden Egg." *Southern Lumberman* 112 (August 25, 1923): 36–38.

———. "A Naval Stores Handbook Dealing with the Production of Pine Gum or Oleoresin." Miscellaneous Publication No. 209. Washington: U.S. Department of Agriculture, January 1935.

Goff, John H. "The Great Pine Barrens." *Emory University Quarterly* 5 (March 1949): 20–31.

Goldenberg, Joseph A. *Shipbuilding in Colonial America.* Charlottesville: University of Virginia Press, 1976.

Gordon, Peter. "Journal of Peter Gordon." *Our First Visit in America: Early Reports from the Colony of Georgia, 1732–1740*, 3–40. Savannah: Beehive Press, 1974.

Graves, Henry Solon. "Present Condition of American Silviculture." *Proceedings of the Society of American Foresters* 3 (October 1908): 29–40.

———. "Protection of Forests from Fire." U.S. Department of Agriculture, Forest Service Bulletin 82, Washington, 1910.

————. "The Selection System." *Proceedings of the Society of American Foresters* 5 (1910): 1–17.

Gray, Lewis C., and Esther K. Thompson. *History of Agriculture in the Southern United States to 1860.* 2 vols. Publication No. 430. Washington: Carnegie Institute of Washington, 1933.

Greeley, William B. *Forests and Men.* New York: Doubleday, 1951.

Greene, S. W. "The Forest That Fire Made." *American Forests* 37 (October 1931): 583–84, 618.

Hall, Basil. *Travels in North America, in the Years 1827 and 1828.* 3 vols. 1829. Reprint, New York: Arno Press, 1974.

Hale, P. M. *The Woods and Timbers of North Carolina.* Raleigh: P. M. Hale, 1883.

Hall, Stephen P., and Dale F. Schweitzer. "A Survey of the Moths, Butterflies, and Grasshoppers of Four Nature Conservancy Preserves in Southeastern North Carolina." Report to the Nature Conservancy, North Carolina Chapter, Raleigh, 1993.

Haller, John S., Jr. "Sampson of the Terebinthinates: Medical History of Turpentine." *Southern Medical Journal* 77 (June 1984): 750–54.

Hardesty, J. L. "Policy and Process: Ecosystem Management on Department of Defense Lands in North Florida." In *Environmental Policy and Biodiversity*, edited by R. E. Grumbine, 311–26. Washington, D.C.: Island Press, 1994.

Hare, Robert C. "Contribution of Bark to Fire Resistance." *Journal of American Forestry* 63 (April 1965): 248–51.

Harper, Roland M. "A Defense of Forest Fires." *Literary Digest* 47 (August 9, 1913): 208.

————. "Historical Notes on the Relation of Fires to Forests." *Proceedings of the Tall Timbers Fire Ecology Conference* 1 (1962): 10–29.

Harrison, W. H. *How to Get Rich in the South.* Chicago: W. H. Harrison Jr., 1888.

Hayes, Martha Green. "General History of the Turpentine Industry." Unpublished manuscript. Georgia Agrirama, State Museum of Agriculture, Tifton, 1982.

Haywood, J. David, Alton Martin Jr., Finis Harris, and Michael Elliott-Smith. "Restoration of Native Plant Communities in Longleaf Landscapes on the Kisatchie National Forest." *Proceedings of the First Longleaf Alliance Conference.* Longleaf Alliance Report No. 1 (September 1996): 93–95.

Haywood, J. David, Michael Elliott-Smith, Finis Harris, and Alton Martin Jr. "Protecting and Restoring Longleaf Pine Forests on the Kisatchie National Forest in Louisiana." *Proceedings of the Third Longleaf Alliance Regional Conference.* Longleaf Alliance Report No. 5 (July 2001): 133–35.

Herriot, Edwin. "The Manufacture of Turpentine in the South." *DeBows Commercial Review of the South and West* 8 (May 1850): 450–56.

Herty, Charles. "White Paper from Young Southern Pines." *Paper Trade Journal* (March 30, 1933): 23–27.

Hickman, Nollie W. "Black Labor in Forest Industries." In *Mississippi's Piney Woods: A Human Perspective*, edited by Noel Polk, 79–91. Jackson: University Press of Mississippi, 1986.

———. "Logging and Rafting Timber in South Mississippi, 1840–1910." *Journal of Mississippi History* 19 (July 1957): 154–72.

———. *Mississippi Harvest: Lumbering in the Longleaf Pine Belt, 1840–1915*. University, Miss: University of Mississippi, 1962.

———. "The Yellow Pine Industries in St. Tammany, Tangipahoa, and Washington Parishes, 1840–1915." *Louisiana Studies* 5 (Summer 1966): 75–88.

Hillyard, M. B. *The New South: A Description of the Southern States, Noting Each State Separately, and Giving Their Distinctive Features and Most Salient Characteristics*. Baltimore: Manufacturers' Record Co., 1887.

Hood, David Foard, and Laura A. Phillips. National Historic Landmark Nomination, Pinehurst Historic District, 1995.

Hooper, Robert G. "Longleaf Pines Used for Cavities by Red-Cockaded Woodpeckers." *Journal of Wildlife Management* 52 (1988): 392–98.

Hooper, Robert G., and Richard F. Harlow. "Forest Stands Selected by Foraging RCWs." Southeastern Forest Experiment Station, Research Paper SE-259, 1986.

Hooper, Robert G., Dennis L. Krusac, and Danny L. Carlson. "An Increase in a Population of Red-Cockaded Woodpeckers." *Wildlife Society Bulletin* 19 (Fall 1991): 277–86.

Hooper, Robert G., Michael R. Lennartz, and H. David Muse. "Heart Rot and Cavity Tree Selection by Red-Cockaded Woodpeckers." *Journal of Wildlife Management* 55 (February 1991): 323–27.

Hooper, Robert G., and Colin J. McAdie. "Hurricanes and the Long-Term Management of the Red-Cockaded Woodpecker." Typed manuscript, Proceedings of the Third Red-Cockaded Woodpecker Symposium, August 1995.

Hooper, Robert G., Andrew F. Robinson Jr., and Jerome A. Jackson. "The Red-Cockaded Woodpecker: Notes on Life History and Management." U.S. Department of Agriculture, Forest Service, General Report SA-GR 9, March 1980.

Hooper, Robert G., J. Craig Watson, and Ronald E. F. Escano. "Hurricane Hugo's Initial Effects on Red-Cockaded Woodpeckers in the Francis Marion National Forest." *Transactions: Fifty-fifth North American Wildlife and Natural Resources Conference* (1990): 220–24.

Hough, Franklin B. *Report upon Forestry*. 4 vols. Washington: Government Printing Office, 1878, 1880, 1882, 1884.

Hudson, Charles. "The Genesis of Georgia's Indians." *Journal of Southwest Georgia History* 3 (Fall 1985): 1–16.

———. *Knights of Spain, Warriors of the Sun: Hernando de Soto and the South's Ancient Chiefdoms*. Athens: University of Georgia Press, 1997.

Hutchins, John G. *The American Maritime Industries and Public Policy, 1789–1914: An Economic History*. Cambridge: Harvard University Press, 1941.

"Incentives for Endangered Species Conservation: Opportunities in the Sandhills of North Carolina." Environmental Defense Fund, 1995.

Ivy, Thomas Parker. *The Long Leaf Pine*. Southern Pines: Sandhills Citizen Print, 1923.

Jackson, Dale R., and Eric G. Milstrey. "The Fauna of Gopher Tortoise Burrows." In *Gopher Tortoise Relocation Symposium Proceedings*, edited by Joan E. Diemer et al.,

State of Florida Game and Fresh Water Fish Commission, Nongame Wildlife Program Technical Report 5 (1989): 86–98.

Jackson, J. A. "Red-Cockaded Woodpecker *(Picoides borealis)*." In *The Birds of North America*, No. 85, edited by A. Poole and F. Gill. Philadelphia: Academy of Natural Sciences, 1994.

Jahoda, Gloria. *The Other Florida.* 1967. Reprint, New York: Charles Scribner's Sons, 1978.

James, Fran C., Ronald E. F. Escano, Ralph Costa, and Jeffrey R. Walters. "Panel Discussion: Managed Longleaf Pine Forests and Red-Cockaded Woodpeckers." In *Proceedings of the Tall Timbers Fire Ecology Conference* 18 (1993): 371–83.

Janzen, Daniel H. "Seed Predation by Animals." *Annual Review of Ecology and Systematics* 2 (November 1971): 465–92.

Jennings, W. S. "Turpentine Operators' Association Holds Its First Annual Convention." *Florida Times-Union and Citizen,* September 11, 1902.

Johnson, F. Roy. *Riverboating in Lower Carolina.* Murfreesboro: Johnson Publishing Co., 1977.

Johnson, Rhett, and Dean Gjerstad. "Restoring the Longleaf Pine Forest Ecosystem." *Alabama's Treasured Forests* 18 (Fall 1999): 18–19.

Jones, Bill M. *Health-Seekers in the Southwest, 1817–1900.* Norman: University of Oklahoma Press, 1967.

Jones, Frank Morton. "Pitcher-Plant Insects—I." *Entomological News* 15 (January 1904): 14–17.

———. "Pitcher-Plant Insects—II." *Entomological News* 18 (December 1907): 413–20.

———. "Pitcher-Plant Insects—III." *Entomological News* 19 (April 1908): 150–56.

———. "Pitcher-Plants and Their Moths: The Influence of Insect-Trapping Plants on Their Insect Associates." *Natural History* 21 (May–June 1921): 296–316.

Jordan, Terry. *North American Cattle-Ranching Frontiers: Origins, Diffusion, and Differentiation.* Albuquerque: University of New Mexico Press, 1993.

———. *Trails to Texas: Southern Roots of Western Cattle Ranching.* Lincoln: University of Nebraska Press, 1981.

"Journal of a French Traveller in the Colonies, 1965." *American Historical Review* 26 (July 1921): 726–47.

Kaczor, Sue A., and David C. Hartnett. "Gopher Tortoise *(Gopherus polyphemus)* Effects on Soils and Vegetation in a Florida Sandhill Community." *American Midland Naturalist* 123 (January 1990): 100–111.

Kellert, Stephen R., and F. Herbert Bormann. "Closing the Circle." In F. Herbert Bormann and Stephen R. Kellert, eds., *Ecology, Economics, Ethics: The Broken Circle,* 205–10. New Haven: Yale University Press, 1991.

Kelton, Paul. "The Great Southeastern Smallpox Epidemic, 1696–1700." In *Transformation of Southeastern Indians, 1540–1760,* edited by Robbie Ethridge and Charles Hudson, 21–38. Jackson: University Press of Mississippi, 2002.

King, Edward. *The Great South: A Record of Journeys.* Hartford, Conn.: American Publishing Co., 1879.

King, J. Crawford. "The Closing of the Southern Range: An Exploratory Study." *Journal of Southern History* 48 (February 1982): 53–70.

Kirby, Jack Temple. *Poquosin: A Study of Rural Landscape and Society.* Chapel Hill: University of North Carolina Press, 1995.

Kirke, Edmund (James Robert Gilmore). *Among the Pines, or The South in Secession Time.* New York: Gilmore, 1862.

————. *My Southern Friends.* New York: Carleton, 1865.

Kohm, Kathryn A., and Jerry F. Franklin, eds. *Creating a Forestry for the 21st Century: The Science of Ecosystem Management.* Washington: Island Press, 1997.

Komarek, E. V., Sr. "Fire and Animal Behavior." *Proceedings of the Tall Timbers Fire Ecology Conference* 9 (1969): 161–207.

————. "Fire Ecology." *Proceedings of the Tall Timbers Fire Ecology Conference* 1 (1962): 95–107.

————. "Fire Ecology—Grasslands and Man." *Proceedings of the Tall Timbers Fire Ecology Conference* 4 (1965): 169–220.

————. "History of Prescribed Fire and Controlled Burning in Wildlife Management in the South." N.d. Paper in Tall Timbers Fire Ecology File, Tall Timbers Research Station, Tallahassee, Fla.

————. "The Natural History of Lightning." *Proceedings of the Tall Timbers Fire Ecology Conference* 3 (1964): 139–83.

————. "A Quest for Ecological Understanding." *Tall Timbers RS Miscellaneous Publication* 5 (1997): 15–21.

————. "The Role of the Hunting Plantation in the Development of Game Fire Ecology and Management." *Proceedings of the Tall Timbers Fire Ecology Conference* 16 (1979): 167–88.

————. "The Use of Fire: An Historical Background." *Proceedings of the Tall Timbers Fire Ecology Conference* 1 (1962): 7–10.

Kulhavy, David L., Robert G. Hooper, and Ralph Costa, *Red-Cockaded Woodpecker: Recovery, Ecology, and Management.* Nacogdoches, Tex.: Center for Applied Studies in Forestry, College of Forestry, Stephen F. Austin State University, 1995.

Landers, J. Larry. "Disturbance Influences on Pine Traits in the Southeastern United States." *Proceedings of the Tall Timbers Fire Ecology Conference* 17 (1991): 61–98.

Landers, J. Larry, Nathan A. Byrd, and Roy Komarek. "A Holistic Approach to Managing Longleaf Pine Communities." In *Proceedings of the Symposium on the Management of Longleaf Pine,* edited by Robert M. Farrar, 135–67. U.S. Department of Agriculture, Forest Service, General Technical Report SO-75, April 1989.

Landers, J. Larry, David H. Van Lear, and William D. Boyer. "The Longleaf Pine Forests of the Southeast: Requiem or Renaissance?" *Journal of Forestry* 93 (November 1995): 39–44.

Lane, Ferdinand C. *The Story of Trees.* New York: Doubleday, 1953.

Lane, Mills, ed. *General Oglethorpe's Georgia: Colonial Letters, 1733–1743.* 2 vols. Savannah: Beehive Press, 1975.

Lee, David S., and James F. Parnell, eds. "Endangered, Threatened, and Rare Fauna of

North Carolina: Part III, A Re-evaluation of the Birds." North Carolina Biological
  Survey and the North Carolina State Museum of Natural Sciences, Raleigh, 1990.
Lefler, Hugh Talmadge, and Albert Ray Newsome. *North Carolina: The History of a
  Southern State*. Chapel Hill: University of North Carolina Press, 1973.
Lennartz, Michael R. "The Red-Cockaded Woodpecker: Old-Growth Species in a
  Second-Growth Landscape." *Natural Areas Journal* 8 (July 1988): 160–65.
———. "Reply to Edward C. Fritz," *Natural Areas Journal* 9 (January 1989): 4.
Lieberman, Paul, and Chester Goolrick. "Naval Stores Ages-Old, but Few Like Living
  in Past." *Atlanta Journal and Constitution*, December 2, 1979.
Ligon, J. David, Peter B. Stacey, Richard N. Conner, Carl E. Bock, and Curtis S.
  Adkisson. "Report of the American Ornithologists' Union Committee for the
  Conservation of the Red-Cockaded Woodpecker." *The Auk* 103 (October 1986):
  848–85.
"The Lime Sink Region." *New York Lumber Trade Journal*, June 15, 1887.
Lindeman, Stephen T. "An Evaluation of Ecological Forest Management for the Red
  Hills Region of Georgia and Florida." M.A. thesis, Duke University, 1994.
Long, Ellen Call. "Notes of Some of the Forest Features of Florida, with Items of
  Tree Growth in That State." *Proceedings of the American Forestry Congress*
  (December 1888): 38–41.
Ludwig, Donald, Ray Hilborn, and Carl Walters. "Uncertainty, Resource
  Exploitation, and Conservation: Lessons from History." *Science* 260 (April 1993):
  17, 36.
Lyell, Charles. *A Second Visit to the United States of North America*. Vol. 1. New York:
  Harper and Brothers, 1849.
———. *Travels in North America in the Years 1841–2*. 1845. Reprint, New York: Arno
  Press, 1978.
MacLeod, John. "The Tar and Turpentine Business of North Carolina." *Monthly
  Journal of Agriculture* 2 (July 1846): 13–19.
"The Manufacture of Turpentine in the South." *Commercial Review of the South and
  West* 8, Old Series (June 1850): 450–56.
Martin, William H., and Stephen G. Boyce. "Introduction: The Southeastern Setting."
  In *Biodiversity of the Southeastern United States/Lowland Terrestrial Communities*,
  edited by William H. Martin, Stephen G. Boyce, and Arthur C. Echternacht, 1–46.
  New York: Wiley, 1993.
Matsch, Charles L. *North America and the Great Ice Age*. New York: McGraw-Hill, 1976.
Mattoon, Wilbur R. "Longleaf Pine." U.S. Department of Agriculture, Forest Service,
  Bulletin No. 1061 (July 1922; rev. August 1925).
———. "Slash Pine—An Important Second-Growth Tree." *Proceedings of the Society of
  American Foresters* 11 (July 1916): 405–16.
Maxwell, Robert S., and Robert D. Baker. *Sawdust Empire: The Texas Lumber Industry,
  1830–1940*. College Station: Texas A & M University Press, 1983.
Mayor, Archer H. *Southern Timberman: The Legacy of William Buchanan*. Athens:
  University of Georgia Press, 1988.

McDaniel, Sidney. "Vegetation of the Piney Woods." In *Mississippi's Piney Woods: A Human Perspective*, edited by Noel Polk, 173–82. Jackson: University Press of Mississippi, 1986.

McFarlane, Robert W. *A Stillness in the Pines: The Ecology of the Red-Cockaded Woodpecker*. New York: Norton, 1992.

McWhiney, Grady, and Forrest McDonald, "Celtic Origins of Southern Herding Practices." *Journal of Southern History* 51 (May 1985): 165–82.

McWhite, Richard W., Dana R. Green, Carl J. Petrick, and Stephen M. Seiber. *Natural Resources Management Plan, Eglin Air Force Base, Florida, for Plan Period March 1993– March 1997*. N.p.: Department of the Air Force, 1993.

Means, D. Bruce. "Impacts on Diversity of the 1985 Land and Resource Management Plan for National Forests in Florida." A Report to the Wilderness Society, 1987.

Means, D. Bruce, and Gerald Grow. "The Endangered Longleaf Pine Community." *Enfo* (September 1985): 1–12.

Merrens, Harry Roy. *Colonial North Carolina in the Eighteenth Century: A Study in Historical Geography*. Chapel Hill: University of North Carolina Press, 1964.

Milanich, Jerald T. *Laboring in the Fields of the Lord: Spanish Missions and Southeastern Indians*. Washington: Smithsonian Institution Press, 1999.

Millet, Donald J. "The Lumber Industry of 'Imperial' Calcasieu, 1865–1900." *Louisiana History* 7 (Winter 1966): 51–69.

Mirov, Nicholas T. *The Genus Pinus*. New York: Ronald Press Co., 1967.

Mirov, Nicholas T., and Jean Hasbrouck. *The Story of Pines*. Bloomington: Indiana University Press, 1976.

Mitchell, William R. *Landmarks: The Architecture of Thomasville and Thomas County, Ga., 1820–1980*. Thomasville: Thomasville Landmarks, Inc., 1980.

Mohr, Charles. "The Long-Leaved Pine." *Garden and Forest* (July 25, 1888): 261–62.

———. "Present Condition of Forests of Longleaf Pine in Alabama and Mississippi." *The Forester* 4 (July 1898): 146–48.

———. *The Timber Pines of the Southern United States*. U.S. Department of Agriculture, Division of Forestry, Bulletin No. 13, 1897.

Moore, Francis. "A Voyage to Georgia, Begun in the Year 1735." In *Our First Visit in America: Early Reports from the Colony of Georgia, 1732–1740*, 81–158. Savannah: Beehive Press, 1974.

Morrison, J. S. *Greek Oared Ships, 900–322 B.C.* London: Cambridge University Press, 1968.

Muir, John. *A Thousand-Mile Walk to the Gulf*. Boston: Houghton Mifflin, 1916.

Murphy, Louis B. "Forest Taxation Experience of the States and Conclusions Based on Them." *Journal of Forestry* 22 (May 1924): 453–63.

Mutch, Robert W. "Wildland Fires and Ecosystems—A Hypothesis." *Ecology* 51 (Autumn 1970): 1046–51.

Myers, Ronald L. "Scrub and High Pine." In *Ecosystems of Florida*, edited by Ronald L. Myers and John J. Ewel, 150–93. Orlando: University of Central Florida Press, 1990.

Napier, John Hawkins, III. *Lower Pearl River's Piney Woods: Its Land and People*.

University, Miss.: University of Mississippi Center for the Study of Southern Culture, 1985.

———. "Piney Woods Past: A Pastoral Elegy." In *Mississippi's Piney Woods: A Human Perspective*, edited by Noel Polk, 12–24. Jackson: University Press of Mississippi, 1986.

"Natural Resources Management Plan," Eglin Air Force Base, 1993–97, March 1993.

The Nature Conservancy and North Carolina Natural Heritage Program. *Rare and Endangered Plant Survey and Natural Area Inventory for Fort Bragg and Camp MacKall Military Reservations, North Carolina*. Carrboro and Raleigh: The Nature Conservancy, 1993.

"New Source of Wealth in South's Pine Forests." *Manufacturers' Record* 101 (July 1932): 13–14.

Newton, M. B., Jr. "Water-Powered Sawmills and Related Structures in the Piney Woods." In *Mississippi's Piney Woods: A Human Perspective*, edited by Noel Polk, 155–72. Jackson: University Press of Mississippi, 1986.

Nonnemacher, Robert M. *Impacts of Forest Industries on Forest Resources in the South*. Miscellaneous Publication No. 1473. Washington: U.S. Department of Agriculture, 1989.

Norse, Elliott A., ed. *Ancient Forests of the Pacific Northwest*. Washington, D.C.: Island Press, 1990.

Noss, Reed F. "The Longleaf Pine Landscape of the Southeast: Almost Gone and Almost Forgotten." *Endangered Species Update* 5 (1988): 1–8.

———. "Longleaf Pine and Wiregrass: Keystone Components of an Endangered Ecosystem." *Natural Areas Journal* 9 (October 1989): 211–13.

———. "Sustainable Forestry or Sustainable Forests?" In *Defining Sustainable Forestry*, edited by Gregory H. Aplet, Nels Johnson, Jeffrey T. Olson, and V. Alaric Sample, 17–42. Washington, D.C.: Island Press, 1993.

Noss, Reed F., and Allen Y. Cooperrider. *Saving Nature's Legacy: Protecting and Restoring Biodiversity*. Washington: Island Press, 1994.

Noss, Reed F., Edward T. LaRoe III, and J. Michael Scott. "Endangered Ecosystems of the United States: A Preliminary Assessment of Loss and Degradation." U.S. Department of the Interior, National Biological Service, Biological Report 28, February 1995.

Oden, Jack P. "Charles Holmes Herty and the Birth of the Southern Newsprint Paper Industry, 1927–1940." *Journal of Forest History* 21 (April 1977): 77–89.

Oglethorpe, James, to the Trustees, February 10, 20, 1733. In John Percival, the Earl of Egmont Papers, Phillips Collection, Hargrett Rare Book and Manuscript Library, University of Georgia Libraries, 14200: 34–35. Citation from National Park Service Web Site.

Olmsted, Frederick Law. *A Journey in the Seaboard Slave States*. New York: Dix and Edwards, 1856.

Outcalt, Kenneth W., and Raymond M. Sheffield. "The Longleaf Pine Forest: Trends and Current Conditions." U.S. Department of Agriculture, Forest Service, Southern Research Station Resource Bulletin SRS-9, 1996.

Outland, Robert Boone, III. "Servants of the Turpentine Orchards: Laborers in the Southeastern North Carolina Naval Stores Industry, 1835–1860." M.A. thesis, Appalachian State University, 1991.

Owsley, Frank L. *Plain Folk of the Old South*. Baton Rouge: Louisiana State University Press, 1949.

Padgett, James A., ed. "With Sherman through Georgia and the Carolinas: Letters of a Federal Soldier," Part II. *Georgia Historical Quarterly* 33 (March 1949): 49–81.

Paisley, Clifton. *From Cotton to Quail: An Agricultural Chronicle of Leon County, Florida, 1860–1967*. Tallahassee: University Presses of Florida, 1968.

Pederson, Neil, Leon Neel, and John Kush. "Ecosystem Management Ideas for the Longleaf Pine Ecosystem." *Proceedings of the First Longleaf Alliance Conference* 1 (1996): 135.

Peet, Robert K., et al. "Mechanisms of Co-Existence in Species-Rich Grassland." Poster, Ecological Society of America, 1990.

Peet, Robert K., and Dorothy J. Allard. "Longleaf Pine Vegetation of the Southern Atlantic and Eastern Gulf Coast Regions: A Preliminary Classification." *Proceedings of the Tall Timbers Fire Ecology Conference* 18 (1993): 45–82.

Perry, G. W. *A Treatise on Turpentine Farming: Being a Review of Natural and Artificial Obstructions, with Their Results in Which Many Erroneous Ideas Are Exploded; with Remarks on the Best Method of Making Turpentine*. New Bern: N.p., 1859.

Perry, Percival. "Naval Stores." In *Encyclopedia of Southern Culture*, edited by Charles Reagan Wilson and William Ferris, 40. Chapel Hill: University of North Carolina Press, 1989.

———. "The Naval Stores Industry in the Ante-Bellum South, 1789–1861." Ph.D. diss., Duke University, 1947.

———. "The Naval-Stores Industry in the Old South, 1790–1860." *Journal of Southern History* 34 (November 1968): 509–26.

Pikl, Ignatz James, Jr. "Pulp and Paper and Georgia: The Newsprint Paradox." *Forest History* 12 (October 1968): 6–19.

Pinchot, Gifford. *The Training of a Forester*. Philadelphia: Lippincott, 1914.

Pinchot, Gifford, and W. W. Ashe. *Timber Trees and Forests of North Carolina*. Bulletin No. 6, North Carolina Geological Survey. Winston: M. I. and J. C. Stewart, Public Printers, 1897.

"The Pine Forests of the South." *DeBows Review* 3—After the War Series (February 1867): 196–98.

Pinehurst and the Village Chapel. Pinehurst, N.C.: Pinehurst Religious Association, 1957.

"The Piney Woods." *DeBows Review* 21, Old Series (July 1861): 361–69.

Piney Woods Hotel, Thomasville, Georgia (1886). The Tourist Committee of the Thomasville–Thomas County Chamber of Commerce, 1967.

Platt, William J., Gregory W. Evans, and Mary M. Davis. "Effects of Fire Season on Flowering of Forbs and Shrubs in Longleaf Pine Forests." *Oecologia* 76, no. 3 (1988): 353–63.

Platt, William J., Gregory W. Evans, and Stephen L. Rathbun. "The Population

Dynamics of a Long-Lived Conifer (*Pinus Palustris*)." *American Naturalist* 131 (April 1988): 491–525.

Platt, William J., Jeff S. Glitzenstein, and Donna R. Streng. "Evaluating Pyrogenicity and Its Effects on Vegetation in Longleaf Pine Savannas." *Proceedings of the Tall Timbers Fire Ecology Conference* 17 (May 1989): 143–62.

Pooler, Franklin. Letter to District Forester, Albuquerque, New Mexico, April 29, 1911. Typed manuscript in Eglin Air Force Base Library.

Porcher, Francis Peyre. "Uses of Rosin and Turpentine in Old Plantation Days." In *Naval Stores—History, Production, Distribution, and Consumption*, edited by Thomas Gamble, 29–30. Savannah: N.p., n.d.

Powell, William S. "What's in a Name? Why We're All Called Tar Heels." *Tar Heel* 10 (March 1982). Reprint.

Probst, John R., and Thomas R. Crow. "Integrating Biological Diversity and Resource Management: An Essential Approach to Productive, Sustainable Ecosystems." *Journal of Forestry* 89 (February 1991): 12–17.

Pyne, Stephen J. *Fire in America: A Cultural History of Wildland and Rural Fire.* Princeton, N.J.: Princeton University Press, 1982.

Quarterman, Elsie, and Catherine Keever. "Southern Mixed Hardwood Forest: Climax in the Southeastern Coastal Plain, U.S.A." *Ecological Monographs* 32 (Spring 1962): 167–85.

Recknagel, A. B. "Certain Limitations of Forest Management." *Proceedings of the Society of American Foresters* 8 (July 1913): 227–31.

Record of Decision: Final Environmental Impact Statement for the Management of the Red-Cockaded Woodpecker and Its Habitat on National Forests in the Southern Region. U.S. Department of Agriculture, Forest Service Southern Region, Management Bulletin R8-MB 73, 1995.

Reed, Gerry. "Saving the Naval Stores Industry: Charles Holmes Herty's Cup-and-Gutter Experiment, 1900–1905." *Journal of Forest History* 26 (October 1982): 168–75.

Rees, Abraham. *Rees's Naval Architecture* (1819–20). N.p.: David and Charles Reprints, n.d.

Report of the Secretary of Agriculture, 1892. Washington: Government Printing Office, 1893.

Richards, Horace G., and Sheldon Hudson. "The Atlantic Coastal Plain and the Appalachian Highlands in the Quaternary." In *The Quaternary of the United States*, edited by H. E. Wright Jr. and D. G. Frey, 129–36. Princeton, N.J.: Princeton University Press, 1965.

Richardson, Emma G., and Thomas C. Richardson. *History of Aberdeen.* Aberdeen, N.C.: Malcolm Blue Historical Society, 1976.

Robbins, Louise E., and Ronald L. Myers. "Seasonal Effects of Prescribed Burning: A Review." Tall Timbers Research, Inc., Miscellaneous Publication No. 8, 1992.

Robbins, William G. *Lumberjacks and Legislators: Political Economy of the U.S. Lumber Industry, 1890–1941.* College Station: Texas A & M University Press, 1982.

Roberts, E. V. "Management Plan [for] Choctawhatchee National Forest, 1931 Revision." Eglin Air Force Base Archives.

Robinson, Gordon. *The Forest and the Trees: A Guide to Excellent Forestry*. Washington: Island Press, 1988.

Rogers, William Warren. *Foshalee: Quail Country Plantation*. Tallahassee, Fla.: Sentry Press, 1989.

Romans, Bernard. *A Concise Natural History of East and West Florida, 1775*. Edited by Kathryn E. Holland Braund. Tuscaloosa: University of Alabama Press, 1999.

Rosenkrantz, Barbara Gutmann, ed. *From Consumption to Tuberculosis: A Documentary History*. New York: Garland Publishing, 1994.

Rostlund, Erhard. "The Myth of the Natural Prairie Belt in Alabama." *Annals of the Association of American Geographers* 47 (December 1957): 392–411.

Roth, Filibert. "Notes on the Structure of the Wood of Five Southern Pines." In *The Timber Pines of the Southern United States*, by Charles Mohr, 141–68. U.S. Department of Agriculture, Division of Forestry, Bulletin No. 13, 1897.

Rudolph, D. C., R. N. Conner, and Janet Turner. "Competition for Red-Cockaded Woodpecker Roost and Nest Cavities: Effects of Resin Age and Entrance Diameter." *Wilson Bulletin* 102 (January–March 1990): 23–36.

Ruffin, Edmund. *Agricultural, Geological, and Descriptive Sketches of Lower North Carolina, and the Similar Adjacent Lands*. Raleigh, 1861.

———. *The Diary of Edmund Ruffin*. Edited by William Kauffman Scarborough. Baton Rouge: Louisiana State University Press, 1972.

———. "Notes of a Steam Journey." *Farmers' Register* (1840): 243–52.

Ryan, Frank, M.D. *The Forgotten Plague: How the Battle against Tuberculosis Was Won — and Lost*. Boston: Little, Brown, 1992.

Saley, Met L. "Relation of Forestry to the Lumbering Industry." *Proceedings of the American Forestry Association* 10 (1894): 147–50.

Sargent, Charles S. "Report on the Forests of North America (Exclusive of Mexico)." In U.S. Bureau of the Census, *Tenth Census of the United States*. Vol. 9. Washington: Government Printing Office, 1884.

Sargent, Charles S. *Sylva of North America*. Vol. 11, *Coniferae*. Boston: Houghton Mifflin, 1897.

Scammon, Charles Melville. *Marine Mammals of the Northwestern Coast of North America, Described and Illustrated: Together with an Account of the American Whale-Fishery*. San Francisco: J. H. Carmany, 1874.

Schafale, Michael, and Alan Weakley. *Classification of the Natural Communities of North Carolina: Third Approximation*. Raleigh: North Carolina Natural Heritage Program, 1990.

———. "Ecological Concerns about Pine Straw Raking in the Southeastern Longleaf Pine Ecosystem." *Natural Areas Journal* 10 (October 1990): 220–21.

Scharitz, Rebecca R., L. R. Boring, D. H. Van Lear, and J. E. Pinder III. "Integrating Ecological Concepts with Natural Resource Management of Southern Forests." *Ecological Applications* 3 (August 1992): 226–37.

Schaw, Janet. *Journal of a Lady of Quality*. Edited by Evangeline Walker Andrews. New Haven: Yale University Press, 1939.

Schiff, Ashley. *Fire and Water: Scientific Heresy in the U.S. Forest Service*. Cambridge: Harvard University Press, 1962.

Schoepf, Johann David. *Travels in the Confederation*. Translated and edited by Alfred J. Morrison. Philadelphia: William J. Campbell, 1911.

Schorger, A. W., and H. S. Betts. "The Naval Stores Industry." U.S. Department of Agriculture Bulletin 229, July 1915.

Schwarz, G. Frederick. *The Longleaf Pine in Virgin Forest: A Silvical Study*. New York: Wiley, 1907.

Sharrer, G. Terry. "Naval Stores, 1781–1881." In *Material Culture of the Wooden Age*, edited by Brooke Hindle, 241–361. Tarrytown, N.Y.: Sleepy Hollow Press, 1981.

Shelton, Jane Twitty. *Pines and Pioneers: A History of Lowndes County Georgia, 1825–1900*. Atlanta: Cherokee, 1976.

Sherrard, Thomas H. "A Working Plan for Forest Lands in Hampton and Beaufort Counties, South Carolina." U.S. Department of Agriculture, Bureau of Forestry. Bulletin No. 43. Washington: Government Printing Office, 1903.

Shofner, Jerrell H. "Forced Labor in the Florida Forests, 1880–1950." *Journal of Forest History* 25 (January 1981): 14–25.

Sierra Club et al. v. Lyng et al. 694 F. Supp. 1260 (E.D. Tex. 1988), June 17, 1988.

Silver, Timothy. *A New Face on the Countryside: Indians, Colonists, and Slaves in South Atlantic Forests, 1500–1800*. Cambridge: Cambridge University Press, 1990.

Silvertown, Jonathan W. "The Evolutionary Ecology of Mast Seeding in Trees." *Biological Journal of the Linnean Society* 14 (September 1980): 235–50.

Sinclair, Wayne A., Howard H. Lyon, and Warren T. Johnson. *Diseases of Trees and Shrubs*. Ithaca, N.Y.: Cornell University Press, 1987.

Sitton, Thad. *Backwoodsmen: Stockmen and Hunters along a Big Thicket River Valley*. Norman: University of Oklahoma Press, 1995.

Sitton, Thad, and James H. Conrad. *Nameless Towns: Texas Sawmill Communities, 1880–1942*. Austin: University of Texas Press, 1998.

Slaughter, Thomas. *The Natures of John and William Bartram*. New York: Vintage Books, 1996.

Smith, David C. *History of Papermaking in the United States, 1691–1969*. New York: Lockwood, 1970.

Snyder, James R. "'Mutch Ado about Nothing.'" *Oikos* 43 (December 1984): 404.

Somers, Robert. *The Southern States since the War*. London: Macmillan, 1871.

Southern Forest Resource Assessment, <http://www.srs.fs.fed.us/sustain/>.

"The Southern Pine Forests—Turpentine." *DeBows Review* 18 (February 1855): 188–91.

Southern Pines: Better Profits for Marginal Lands. U.S. Department of Agriculture. Washington: Government Printing Office, 1984.

"Southward the Paper-Making Industry Moves." *Manufacturers Record* 97 (March 13, 1930): 56.

Speake, Dan W. "The Gopher Tortoise Burrow Community." In *The Future of Gopher*

Tortoise Habitats, edited by Ren Lohoefner et al., 44–47. Proceedings of the Second Annual Meeting of the Gopher Tortoise Council, 1981.

Speir, Robert F., M.D. *Going South for the Winter with Hints to Consumptives*. New York: E. O. Jenkins, 1873.

The Spirit of Massachusetts: Building a Tall Ship, 1983–1984. Maine: Thorndike Press, 1984.

Spring, Samuel N. "A Report of the Lumbering of Loblolly Pine by the E. P. Burton Lumber Co., South Carolina." M.A. thesis, Yale University, 1902.

Stanley, T. R. "Ecosystem Management and the Arrogance of Humanism." *Conservation Biology* 9 (April 1995): 255–62.

Steen, Harold K. "The Piney Woods: A National Perspective." In *Mississippi's Piney Woods: A Human Perspective*, edited by Noel Polk, 3–11. Jackson: University Press of Mississippi, 1986.

———, *The U.S. Forest Service: A History*. Seattle: University of Washington Press, 1976.

Steuart, William M. "Turpentine and Rosin." In U.S. Bureau of the Census, *Twelfth Census of the United States: Manufacturers* (Part 3), 9:1003–12. Special Reports on Selected Industries. Washington: Government Printing Office, 1902.

Stoddard, Herbert L. *The Bobwhite Quail: Its Habits, Preservation, and Increase*. New York: Charles Scribner's Sons, 1931.

———. *Memoirs of a Naturalist*. Norman: University of Oklahoma Press, 1969.

———. "Some Techniques of Controlled Burning in the Deep Southeast." *Proceedings of the Tall Timbers Fire Ecology Conference* 1 (1962): 133–43.

———. "Use of Fire in Pine Forests and Game Lands of the Deep Southeast." *Proceedings of the Tall Timbers Fire Ecology Conference* 1 (1962): 31–42.

Stokes, George A. "Log-Rafting in Louisiana." *Journal of Geography* 58 (February 1959): 81–89.

———. "Lumbering and Western Louisiana Cultural Landscapes." *Annals of the Association of American Geographers* 47 (September 1957): 250–66.

"The Story of Naval Stores." Compiled by workers of the Florida Writers' Project (WPA). *Florida Highways* 11 (May 1943): 11–15, 35–37; (July 1943): 15–18, 31–35.

Stout, I. Jack, and Wayne R. Marion. "Pine Flatwoods and Xeric Pine Forests of the Southern (Lower) Coastal Plain." In *Biodiversity of the Southeastern United States/ Lowland Terrestrial Communities*, edited by William H. Martin, Stephen G. Boyce, and Arthur C. Echternacht, 373–446. New York: Wiley, 1993.

Surface, Henry E., and Robert E. Cooper. "Suitability of Longleaf Pine for Paper Pulp." U.S. Department of Agriculture, Bulletin 72, 1914.

"Tables of Allowances of Equipments, Outfits, Stores Etc, for Each Class of Vessels in the Navy of the United States." Washington, D.C.: Alexander and Barnard, 1844.

Tanner, James. *The Ivory-Billed Woodpecker*. New York: National Audubon Society, 1942.

Tebo, Mary. "The Southeastern Piney Woods: Describers, Destroyers, Survivors." M.A. thesis, Florida State University, 1985.

"Technology Transfer Plan: Longleaf Pine Management." U.S. Department of Agriculture, Forest Service, 1986.

Tenth Census of the United States, 1880: Forests. Vol. 9. Washington: Government Printing Office, 1884.

Thomasville, Georgia: The Great Winter Resort among the Pines. Thomasville: Thomasville Business League, 1901.

Thomasville (Among the Pines) and Thomas County, Georgia. Thomasville: Triplett and Burr, 1888.

Thompson, Kenneth. "Wilderness and Health in the Nineteenth Century." *Journal of Historical Geography* 2 (1976): 145–61.

Torr, Cecil. *Ancient Ships.* Chicago, Argonaut, 1964.

"A Trip through the Varied and Extensive Operations of the John L. Roper Lumber Co. in Eastern North Carolina and Virginia." *American Lumberman* (April 27, 1907). Weyerhaeuser Co. reprint.

Troyer, James R., *Nature's Champion: B. W. Wells, Tar Heel Ecologist.* Chapel Hill: University of North Carolina Press, 1993.

"Turpentine: Hints for Those about to Engage in Its Manufacture." *DeBows Review* 19 (October 1855): 486–89.

"Turpentine Making." *Carolina Cultivator* 1 (January 1856): 348–50.

U.S. Bureau of the Census. *Twelfth Census of the United States: Manufacturers* (Part 3), Special Reports on Selected Industries. Washington: Government Printing Office, 1908.

U.S. Fish and Wildlife Service. *Recovery Plan for the Red-Cockaded Woodpecker* (Picoides borealis). 2nd rev. Atlanta: U.S. Fish and Wildlife Service, 2003.

U.S. Forest Service. "Southern Research Station Employees Honored for Longleaf Pine Ecosystem Restoration Research." News release, June 5, 2000, Southern Research Station, Asheville, North Carolina.

Vance, Rupert B. *The Human Geography of the South: A Study in Regional Resources and Human Adequacy.* New York: Russell and Russell, 1932; rev. 1935.

Wahlenberg, W. G. "Effect of Fire and Grazing on Soil Properties and the Natural Reproduction of Longleaf Pine." *Journal of Forestry* 33 (March 1935): 331–37.

———. *Longleaf Pine: Its Use, Ecology, Regeneration, Protection, Growth, and Management.* Washington: Charles Lathrop Pack Forestry Foundation, 1946.

Walker, H. Jesse, and James M. Coleman. "Atlantic and Gulf Coastal Plain." In *Geomorphic Systems of North America*, edited by Will L. Graf, 51–110. Boulder, Colo.: Geological Society of America, 1987.

Walker, Jimmy S. "Potential Red-Cockaded Woodpecker Habitat Produced on a Sustained Basis under Different Silvicultural Systems." In *Red-Cockaded Woodpecker: Recovery, Ecology, and Management*, edited by David L. Kulhavy, Robert G. Hooper, and Ralph Costa, 112–30. Nacogdoches, Tex.: Center for Applied Studies, School of Forestry, Stephen F. Austin State University, 1995.

Walker, Joan L. "Rare Vascular Plant Taxa Associated with the Longleaf Pine Ecosystems: Patterns in Taxonomy and Ecology." *Proceedings of the Tall Timbers Fire Ecology Conference* 18 (May–June 1991): 105–26.

————. "Regional Restoration: A Modest Proposal." Abstract. In *Report to Participants: Longleaf Pine Ecosystem Restoration: Toward a Regional Strategy*, 7–9. Washington: U.S. Department of Agriculture, Forest Service, 1995.

Walker, Joan L., and Robert K. Peet. "Composition and Species Diversity of Pine-Wiregrass Savannas of the Green Swamp, North Carolina." *Vegetation* 55 (April 16, 1983): 163–79.

Walters, Carl. *Adaptive Management of Renewable Resources*. New York: Macmillan, 1986.

Walters, Jeffrey R. "Application of Ecological Principles to the Management of Endangered Species: The Case of the Red-Cockaded Woodpecker." *Annual Review of Ecology and Systematics* 22 (1991): 505–23.

Walters, Jeffrey R., Phillip D. Doerr, and J. H. Carter III. "The Cooperative Breeding System of the Red-Cockaded Woodpecker." *Ethology* 78 (1988): 275–305.

Ware, Stewart, Cecil Frost, and Phillip D. Doerr. "Southern Mixed Hardwood Forest: The Former Longleaf Pine Forest." In *Biodiversity of the Southeastern United States/ Lowland Terrestrial Communities*, edited by William H. Martin, Stephen G. Boyce, and Arthur C. Echternacht, 447–93. New York: Wiley, 1993.

Watson, J. Craig, Robert G. Hooper, Danny L. Carlson, William E. Taylor, and Timothy E. Milling. "Restoration of the Red-Cockaded Woodpecker Population on the Francis Marion National Forest: Three Years Post Hugo." In *Red-Cockaded Woodpecker: Recovery, Ecology, and Management*, edited by David L. Kulhavy, Robert G. Hooper, and Ralph Costa, 172–82. Nacogdoches, Tex.: Center for Applied Studies in Forestry, College of Forestry, Stephen F. Austin State University, 1995.

Watts, W. A. "Postglacial and Interglacial Vegetation History of Southern Georgia and Central Florida." *Ecology* 52 (Summer 1971): 676–90.

————. "Vegetational History of the Eastern United States 25,000 to 10,000 Years Ago." In *Late Quaternary Environments*, edited by H. E. Wright Jr., 1:294–310. Minneapolis: University of Minnesota Press, 1983.

Watts, W. A, B. C. S. Hansen, and E. C. Grimm. "Camel Lake: A 40,000-Year Record of Vegetational and Forestry History from North Florida." *Ecology* 73 (June 1992): 1056–66.

Weigl, Peter, Michael A. Steele, Lori J. Sherman, James C. Ha, and Terry L. Sharpe. "The Ecology of the Fox Squirrel (*Sciurus niger*) in North Carolina: Implications for Survival in the Southeast." *Bulletin* 24, Tall Timbers Research Station, 1989.

Wellman, Manly Wade. *The County of Moore, 1847–1947*. Southern Pines, N.C.: Moore County Historical Association, 1962.

Wells, B. W. "Ecological Problems of the Southeastern United States Coastal Plain." *Botanical Review* 8 (October 1942): 533–61.

————. *The Natural Gardens of North Carolina*. 1932. Reprint, Chapel Hill: University of North Carolina Press, 2002.

Wells, B. W., and I. V. Shunk. "The Vegetation and Habitat Factors of the Coarser Sands of the North Carolina Coastal Plain: An Ecological Study." *Ecological Monographs* 1 (October 1931): 465–520.

Wells, Robert W. *"Daylight in the Swamp!"* New York: Doubleday, 1978.

Wesley, John. "An Extract of the Rev. Mr. John Wesley's Journal, 1735–1737." In *Our First Visit in America: Early Reports from the Colony of Georgia, 1732–1740*, 185–242. Savannah: Beehive Press, 1974.

Whelchel, Jasper E. "Lumber and Timber Products." In U.S. Bureau of the Census, *Twelfth Census of the United States: Manufacturers* (Part 3), 9:583–645. Special Reports on Selected Industries. Washington: Government Printing Office, 1908.

Whitefield, George. "A Journal of a Voyage from London to Savannah in Georgia." In *Our First Visit in America: Early Reports from the Colony of Georgia, 1732–1740*, 281–314. Savannah: Beehive Press, 1974.

Whitney, Gordon C. *Coastal Wilderness to Fruited Plain: A History of Environmental Change in Temperate North America, 1500 to the Present.* Cambridge: Cambridge University Press, 1994.

Whittle, C. A. "The South as a Source of Woodpulp." *Manufacturer's Record* 97 (March 26, 1930): 46.

Wicker, Rassie E. *Miscellaneous Ancient Records of Moore County.* Southern Pines, N.C.: Moore County Historical Association, 1971.

Williams, Ida Bell. *History of Tift County.* Macon, Ga.: J. W. Burke Co., 1948.

Williams, Michael. *Americans and Their Forests: A Historical Geography.* Cambridge: Cambridge University Press, 1989.

Williamson, Hugh. *The History of North Carolina.* 1812. 2 vols. Reprint, Spartanburg, S.C.: Reprint Co., 1973.

Wilson, E. O. *The Diversity of Life.* Cambridge: Harvard University Press, 1992.

Worth, John E. "Spanish Missions and the Persistence of Chiefly Powers." In *Transformation of Southeastern Indians, 1540–1760*, edited by Robbie Ethridge and Charles Hudson, 39–64. Jackson: University Press of Mississippi, 2002.

Wright, Gay Goodman. "Turpentining: An Ethno-Historical Study of a Southern Industry and Way of Life." M.A. thesis, University of Georgia, 1979.

Wright, Henry A., and Arthur W. Bailey. *Fire Ecology: United States and Southern Canada.* New York: Wiley, 1982.

Young, Sharon S., and A. P. Mustian. *Impacts of National Forests on the Forest Resources of the South.* Miscellaneous Publication No. 1472. Washington: U.S. Department of Agriculture, 1989.

Index

Koch, Robert, 196
Komarek, Edward, 20, 30

Lake Okeechobee, 1
Land costs, 134
Landscape scale, 254
Lawsuits, 263–64
Leaf River, 1
LeBlond, Richard, 32–37
Les Landes, 147, 178, 199
Lewis and Clark expedition, 11
Lightning, 20, 236
Lightwood, 91, 92, 93
Links Club (New York City), 198
Little St. Juan (Suwannee River), 11
Lizard, six-lined race runner, 51
Loblolly pine, 182–88, 214, 269; names of,
    182; timber quality of, 182, 184; conver-
    sion from longleaf to, 184–85, 186, 209;
    defects of, 212; removal of, 252, 257,
    261
Log pond, 167
Long Leaf, La., 163
Longleaf Alliance, 249, 269
Longleaf pine
—beauty of, 8
—communities of: types of, 32, 44–45;
    composition of, 37
—compared to park, 13
—compared to sea, 13, 16
—cones of, 23, 211
—decline of, 268; compared to Amazon
    rain forest, 2; compared to Douglas fir,
    2; compared to tallgrass prairie, 2; blame
    for, 3; as "social crime," 4; in twentieth
    century, 154, 170–71, 208, 213
—description of, 7, 13–16
—early roads in, 12
—economics of, 216, 218, 220
—ecosystem of, 46–70; biodiversity of, 33,
    37–40; age of, 70, 73; effects of climate
    change on, 73, 79
—extent of, 1, 79
—fire in. See Fire
—forests of: federal ownership of, 160; sales

of, 160–61. See also Southern Homestead
Act
—grass stage of, 24, 25
—health benefits of, 192–96
—income from, 216
—logging of, 155–62; tools for, 155, 161–62.
    See also Lumber industy; Railroads
—longevity of, 8
—lumber of: effects of turpentining on,
    144, 158–59; names for, 152; strength of,
    152; tight grain of, 152; standards of, 155,
    184; N.C. exports of, 157; image problem
    of, 158; uses of, 154–55; waste of, 168–69;
    overproduction of, 169
—management of: disincentives to, 180–81;
    success of, 182, 209–13; goals of, 248–50,
    264; challenges to, 270–71
—monotony of, 14–16, 33, 271
—needles of, 21, 27
—old-growth populations of, 2, 7, 63,
    254–55
—openness of, 75–77
—origins of ranching traditions in, 80–83
—planting of, 211–12, 214–15, 261
—population dynamics of, 235–36
—products of: poles, 216, 220; pulpwood
    and fence posts, 216; sawtimber, 216;
    pine straw, 223–25. See also Lumber in-
    dustry; Naval stores; Pitch; Rosin; Tar;
    Turpentine
—reasons for growing, 215–16, 218
—reproduction of, 179, 199; effects of hogs
    on, 199–200; effects of fire suppression
    on, 200–207. See also Fire suppression;
    Hogs
—resinous properties of. See Resin
—resistance to diseases of, 215
—restoration of, 4, 248, 260, 264
—seed crop of, 211
—seeds of, 23, 25; and seed predators, 25–
    26
—slow growth of, 25
—soils of, 44–45
—taproot growth of, 23, 24
—virgin forests of, 13, 235, 238

Longleaf Pine Ecosystem Restoration
    Team, 249
Loosestrife, rough-leaf, 42
Louisiana: author's trips to, 163–67, 256–64
Lumber industry: cut and run logging,
    3–4, 168, 207; tools of, 155, 161–62; and
    sawmills, 161; annual longleaf cut, 162,
    167; migratory life of, 163, 171; incentives
    to, 168; investigation of, 170; criticism of,
    170–71. *See also* Railroads
Lyng, Richard E., 232

Macon, Ga., 12, 18, 76, 235
Manufacturers' Record, 187
Mattoon, Wilbur, 183
McKinley, William, 197
McQueen, Anne, 3
Michaux, Andre, 154
Millpond Plantation, 190
Milwaukee Public Museum, 198
Mississippi: author's trips to, 213–18
Mississippi Forestry Association, 214
Mississippi River, 1, 9
Mitchell House, 194
Mobile, Ala., 12, 19, 135, 219
Mobile Daily Register, 196
Mohr, Charles, 169, 176, 177, 178, 180, 184,
    200
Morgan, Jim, 222–26
Mouse, Florida, 51
Muir, John, 13, 87
Murder Creek, 219
Mutch, Robert, 27–28
Myers, Ron, 78–79
Myrtle, wax, 36

Narváez, Pánfilo de, 85
National Conservation Priority Area, 269
National Environmental Protection Act, 229
National Forest Management Act, 229
Native Americans, 3, 73–78; populations
    of, 73–74; agriculture of, 74; tribes, 74,
    77; decline of, 76–78; slave trafficking in,
    77; as cattle herders, 78, 81; effects of on
    longleaf, 84, 248; and Spanish, 85

Nature Conservancy, 38, 241, 253, 256
Naval stores, 27, 85, 86; naval uses of, 86–
    90; origin of term, 87; exports of, 90–91;
    markets for, 90–91. *See also* Pitch; Rosin;
    Tar; Turpentine
Naval Stores Conservation Program, 149
Naval Stores Manufacturers Protection
    Association of Georgia, 146
Naval Stores Review, 146
Neel, Leon, 238–41, 247, 250, 258, 263
Nelson, E. W., 198
New Bern, N.C., 143, 157
North Carolina: author's trips to, 32, 34–37,
    62–70, 91–96, 222–26
North Carolina Geological Survey, 145
North Carolina Natural Heritage Program,
    32, 33, 63
North Carolina Wildlife Resources Com-
    mission, 225
North Western Lumberman, 170
Northwest Florida Aquatic Preserves, 253
Northwest Florida Water Management
    District, 253
Noss, Reed, 2, 248, 249, 265
Nottoway River, 98

Oakum, 85, 87, 89
O'Brian, Patrick, 89
Ocmulgee, 76
Oconee River, 98
Oglethorpe, James, 14, 77
Old-growth forests, 2, 176
Olmsted, Frederick Law, 79, 142
Open range, 81–82, 200

Pacific Northwest, 2, 4, 171
Paper mills, 186, 187, 188
Parker, Robert M., Jr., 233–34
Pascagoula River, 19, 98
Patrick, John T., 196
Patterson Tract, 255
Payne's Prairie, 9
Pearl River, 98, 214
PeeDee River, 98
Peet, Robert, 33, 38–40, 45

Sampit River, 98

Sandhill Crane National Wildlife Refuge, 271

Sandhills, 10, 32, 33, 63, 251

Sandhills, North Carolina, 29, 33, 43, 57, 60, 63, 197, 222, 223, 224, 272; as center of biodiversity, 225

Sand pine, 23, 182, 252

Sap, 99. *See also* Resin

Sash saw, 157

Savanna, 2, 10, 11, 13, 32, 38–40, 45, 63, 69, 75–76, 81, 272

Savannah, Ga., 9, 12, 14, 55, 76, 135, 136, 146, 156, 171, 183, 187, 192, 235

Savannah Morning News, 188

Savannah Naval Stores Exchange, 136

Savannah River, 9, 98, 155, 183

Sawmills: before Civil War, 153, 155; on Cape Fear River, 156; number in long-leaf pine range, 158; and sawmill towns, 163–65; description of, 165; output of, 165; operations of, 165–67; workers in, 165–67; noise of, 166; waste at, 169; abandonment of, 171; at Cedar Creek Land and Timber Company, 220

Saw palmetto, 36

Schafale, Michael, 33, 225

Schaw, Janet, 15, 156

Schenck, Carl, 176, 200

Schoepf, Johann David, 81, 195

Schurz, Carl, 176, 200

Schwarz, G. Frederick, 13, 235

Scrape. *See* Turpentining

Scrub oaks, 182

Sedgewick, Walter, 191–92

Sharpe, Terry, 225

Shaw, Aubrey, 91–96

Shortleaf pine, 214

Shunk, I. V., 38

Sierra Club, 230, 264

Sierra Club et al. v. Lyng et al., 230, 232–34

Silver, Timothy, 94

Sinkers, 151

Site preparation, 232

Sitton, Thad, 80

Slade, Leonard, 169

Slash pine (Cuban pine), 36, 178, 182–88; at Eglin Air Force Base, 252–53

Slash Pine Forest Festival, 188

Smith, Latimore, 256

Smith, Matt, 64

Smoke management, 270–71

Smokey Bear, 206

Snake, indigo, 51

Sneeringer, Margaret, 55–57

Society of American Foresters, 206

Society of American Forestry, 212

Soil Bank Program, 188

Sorrie, Bruce, 40

Sousa, John Philip, 196

Southeast: description of, 2; rivers and streams of, 17, 18; geological history of, 17–18; thunderstorms of, 20, 27; rare plants in, 44

Southern Forest Experiment Station, 205

Southern Forest Heritage Museum, 163–67

Southern Forest Resource Assessment, 270

Southern Forestry Congress, 201

Southern Homestead Act, 160

Southern Lumber Manufacturers' Association, 170

Southern Pine Association, 170

Southern Pines, N.C., 196, 271

South River, 92

Sparrow, Bachman's, 31

Spur tracks, 161. *See also* Railroads

Squirrel, gray, 59, 60, 61

Squirrel, southeastern fox, 31, 59–62, 68, 222; description of, 59; size as adaptation of, 59; range of, 60; diet of, 60–61; role in longleaf forest, 62

Squirrel, southern flying, 31, 59

Squirrel, western fox, 59, 60

Staggerbush, 34

Steamboats, 132–33

Steam skidder, 162; effects on forests, 168, 181. *See also* Railroads

Still, 102; effects of copper still on turpentine industry, 102; description of, 103; at

Georgia Agrirama, 103; worm, 103. *See also* Turpentining

Stoddard, Herbert, 198–99, 202–3, 205, 238, 241. *See also* Quail, bobwhite

Stoddard-Neel single-tree selection, 240, 245–46; criticism of, 241, 245

Sullivan, Bo, 64

T. R. Miller Mill Company, 219–221, 221

Talahasochte, 10–11

Tallahassee, Fla., 10, 77, 85, 190, 192

"Tallahassees" (forest openings), 76

Tallapoosa River, 74

Tallgrass prairie, 2, 39, 271

Tall Timbers Research Station, 7, 20, 30, 56, 78, 237; Fire Ecology Conference, 246, 268

Tar, 86, 87, 156; uses of, 87–89, 95; related words, 89; markets for, 90–91; and origin of "Tar Heel," 95–96. *See also* Tar making

Tar making, 87, 90–94; Swedish methods of, 91; kilns, 91–94; amounts of wood used in, 94; charcoal as byproduct of, 94

Tar River, 98

Teredo navalis, 87

Texas Committee on Natural Resources, 230, 264

Texas Committee on Natural Resources v. Berglund, 230

Thinning, 180

Thomas County, Ga., 171, 194

Thomas County Historical Society and Museum, 194

Thomasville, Ga., 7, 37, 55, 171, 190, 238; as sportsman's paradise, 190–92, 197–98; as health resort, 192–96; hotels of, 194–95

Thomasville (Among the Pines), 192

Thomasville Times, 192

Thomasville Times-Enterprise, 196

Thoreau, Henry David, 11

Timber Pines of the Southern United States, 177

Ticks, gopher, 50–51

Tift County, Ga., 108

Tifton, Ga., 102, 103

Titi drains, 36

Tortoise, gopher, 31, 47–54, 69, 217; diet of, 4; status of, 48; range of, 48–49; size of, 49; longevity of, 49, 52; burrows of, 49–50; as keystone species, 51; services of, 51; description of, 52; predators of, 52; effects of urbanization on, 52–53; hunting of, 52–53; as food, 53

"Tree farm," 221

Trent River, 143

Trinity River, 98

Tuberculosis, 193–97; as "greatest killer in history," 193; theories about cause of, 193; remedy for, 193–94; resorts, 194

Turkey, wild, 29, 222

Turkey oak, 33, 34, 35, 252; at Eglin Air Force Base, 254–55

Turpentine, 27; making of, 87, 99–108; export leader, 98; foreign markets for, 98; markets for, 98; monetary value of, 98; N.C. spirits of, 98; products of, 98; transportation of, 98, 131–35; uses of, 100–101, 104–5; inspection of, 136; regulations governing, 136; overproduction of, 146, 179; from stumps, 149. *See also* Turpentine workers; Turpentining

Turpentine Operators' Association, 138, 140

Turpentine workers: in North Carolina, 135, 138; exploitation of, 137

Turpentining: sensory experience of, 97; as migratory industry, 102, 134, 135, 137; fraud in, 136; camps, 137–38; in Florida, 138–39; criticisms of, 139–41; destructive effects of, 141–45; conservative practices, 142, 144–45; nomadic existence of, 143, 144; scrape as product of, 143–44; labor costs of, 145; profits of, 145–46; in France, 147; cup and gutter system, 147–48; at Choctawhatchee National Forest, 178 —methods of: chipping (hacking) trees, 97, 99; cutting boxes, 97, 99; dipping, 97, 99; destructive effects of, 98, 131–49, 207; calendar of, 99–100; North Carolina standards, 135; best methods of, 141–42; crop, 145